四季嘗樂

Seasons' Feasts

Eric Poon 著

U0111350

作者簡介

Eric Poon

Eric Poon 現從事教學工作。

他於 2011 年成立 Funkitchen，教授不同程度的西餐、蛋糕和甜品。

他畢業於英國伯明翰大學，主修廚藝管理學（Culinary Art Management），獲頒榮譽學位。在英國，他曾於地獄廚神 Gordon Ramsay 旗下的餐廳工作。回港後，他曾在香港頂級酒店工作，如香格里拉酒店、文華東方酒店、半島酒店。

Eric Poon 擅長西餐、蛋糕和甜品，曾參與多個媒體電視拍攝及錄影。現時，他是香港西廚學院的廚師導師。

About

Chef Eric Poon

Chef Eric Poon, teaches a variety of cakes, Western cuisines and dessert class tailored to students of different levels, he has set up Funkitchen cooking studio in Taipo in 2011.

Chef Eric Poon graduated from University College Birmingham in England with a Hon(s) degree in Culinary Art Management. He has worked in numerous top Hong Kong Hotels like Shangri-la Hotel, Mandarin Oriental Hong Kong and the Peninsula HK; he also used to work under Gordon Ramsay restaurant in the UK. He is particularly good at making Western cuisines, pastries and desserts, and has been featured in numerous shows and media reports. Currently, he is working as a chef instructor at the Hong Kong Culinary Academy.

代序

4 年前，教室正需要聘用西菜導師，剛好朋友無意中認識了 Eric，便把他介紹給我，我們立即約好飯聚，試了他的手勢，一頓飯便促使他答應了在我們教室當客席導師。

很快，Eric 的課堂便爆滿了，Eric 的烹飪知識豐富，教授認真，無私的分享，令他成為教室最受歡迎的教師之一。

在 Eric 的課堂，除了學到正統的西菜烹飪方法，也學到不少課堂以外的烹飪技巧。他對選材、用料嚴格執著，認真。即使時間不足，也絕不會馬虎了事，連食物的擺盤也做到盡善盡美才滿意下課。

期待 Eric 的新書，透過書本，與大家分享他對西菜的熱誠與研究；透過書本，把一道道色、香、味、美的精緻菜式，展現在大家面前，讓大家在家裏，也能做出一道道五星級精緻菜式。

Foreword

Four years ago, my cookery workshop was looking for an instructor to teach Western cuisine. A friend of mine happened to know Eric and she hooked us up. He then cooked me a meal, and I was convinced instantly that he was the right fit. I was glad that he agreed to be our guest instructor.

Eric's classes fill up very quickly and he's one of the most popular instructors in my workshop because of his profound knowledge in culinary art, his serious attitude when delivering classes and his selfless sharing of all his secret tricks.

Apart from conventional Western cooking skills, Eric also teaches other non-Western methods and techniques. He insists in using the best ingredients and even if he runs out of time, he never settles for anything less than perfect. He only dismisses the class when he's happy with the mise en place.

With his new book, I expect Eric to share his passion and insights on Western cuisine with his readers. I'm sure he would refine every recipe with the best presentation, aroma, flavours and colours, so that his readers could easily re-create fine dining courses in the comfort of their own home.

Rachel Yau

前言

《四季嘗樂》是為每一位喜歡在家煮食、對煮食有熱誠的朋友而寫的。

以春、夏、秋、冬分章節,每一個季節有 3 至 4 個套餐,每一個套餐有三道菜,每道菜都以香港不同當季售賣的材料為基礎。每一個套餐的開首都會和讀者分享設計靈感、事前準備貼士,令讀者更容易明白整個套餐製作的流程,烹調時才會事半功倍。本書也會介紹幾款基本上湯的做法,最後亦會提及新鮮香草的儲存方法。

本書大部份的食譜都可以在我的教室 Funkitchen 學到,在堂上,我除了教授烹飪外,更加喜歡與學生分享有趣的煮食經驗和貼士。曾經有學生問我為甚麼要成為烹飪老師?因為我非常喜歡與人分享烹飪經驗。這促使我從專業廚師轉為烹飪老師。當學生在課堂中有進步而他們亦樂在其中,這都讓我感到欣慰和驕傲,這也是我當教師的原動力。

本書能順利完成,首先要感謝家人的支持、Forms Kitchen 的出版團隊,特別要感謝的是攝影師 Eric Ng,他將菜式拍攝得更加漂亮。

此外,非常感謝亦師亦友的 Rachel Yau,她不只令我出版食譜書的夢想成真,而且不斷提供對於教學和在媒體曝光的意見。

最後,感謝從 2011 年開始在 Funkitchen 上課的學生,他們不斷給我意見,從而激發我的靈感和創意。

Foreword

Seasons' Feasts is designed and written for any home cooks and especially for those who are passionate for foods. It is divided into four seasons, and each season has 3 to 4 menus. Each menu has 3 dishes planned based on the seasonal produces available in Hong Kong. Some thoughts of designing each menu, doing preparation wisely in advance and cooking tips have all clearly mentioned at the beginning of each menu. The readers can easily understand the work flow on how to execute the menu properly. Added to this, some basic stocks used in this book are also written for following and some storage tips for fresh herbs are also mentioned at the end.

Most of the recipes in this book have been taught in my cooking studio 'Funkitchen', where I shared the cooking experiences and tips with my passionate students from all walks of life. I remember one of my students asked me one question, "Why do you want to be a teacher?" I smiled at her and said: "because I love to share with others." Because of this motto, I gave up being a professional chef and turned into a chef instructor. Have seen my students making progress and having a sense of success for their foods they made after my teaching, I am so glad and proud of them. Those are my biggest motives to be a chef teacher and move forward.

Having *Seasons' Feasts* successfully published, I must thank for my family members including my wife and son, who have been giving me support throughout the tough time.

Moreover, thanks the production team from Forms Kitchen, especially my photographer Eric Ng for having such beautiful pictures.

Special thanks to Rachel Yau, my friend and mentor, who has helped me out to have my dream to publish a book come true. She has constantly given me ideas on how to manage the students and chances to expose myself in the media.

Last but not the least, I love my students especially for those who have been continuously given me ideas on what to learn in my lessons, which in turn ignites my passion to continuously persuade my inspiration and creativity in my class menu planning. I love the magic they have created to keep Funkitchen's business running since 2011.

Eric Poon

目錄 CONTENTS

SET/01

spring menu

香 煎 吞 拿 魚
配 柚 子 西 椒 莎 莎 醬

Seared sesame tuna
with yuzu and pepper salsa

構 思 HOW IS IT DESIGNED ?

我很喜歡烹調法國春雞，雖然牠的體型較小，但肉質軟滑，而且味道不太濃烈，非常適合配上香濃樸實的牛肝菌汁；甘筍的甜味可以平衡牛肝菌的濃味，讓整道菜的味道非常和諧。既然主菜味濃，頭盤的味道宜比較清淡，所以我以香煎吞拿魚配柚子西椒莎莎醬；甜點是草莓慕絲蜂蜜脆餅，草莓的濃郁香氣、慕絲的細膩口感，讓你能圓滿享受這美味的一頓。

The French spring chicken, it is small in size but its meat is so tender and light in flavour, is one of my favourites. It is perfectly matched with something creamy and earthy like ceps mushroom cream sauce. Meanwhile the sweetness from the carrot cuts off the heaviness from the cream, which in turn harmonizes the whole dish. The main course is strong in flavour, it needs something light and refreshing to start with. Seared sesame tuna with yuzu and pepper salsa is one of the appropriate dishes that can push up the appetite for the next course. Lastly, the sharpness and creaminess from the strawberry mousse rounds up this whole menu.

燒 法 國 春 雞
伴 牛 肝 菌 蜂 蜜 甘 筍

Oven roasted French spring chicken
with honey roasted carrot and ceps sauce

草 莓 慕 絲
蜂 蜜 脆 餅

Mille-feuille of strawberry mousse
with honey tuile

預 先 準 備　TIPS FOR SMART PREPARATION

+ 柚子西椒莎莎醬可以預先弄好，放入雪櫃冷藏。．

+ 草莓慕絲和牛肝菌汁可以預早一天弄好，放入雪櫃冷藏保鮮。

+ 蜂蜜脆餅亦可以預先做好，放入密封瓶內可保持鬆脆 3 天。

+ 烹調這餐單當天，宜先燒春雞，燒春雞的同時可以準備吞拿魚、蜂蜜甘筍和甜點。

+ The yuzu and pepper salsa can be made in advance and let them macerate in the fridge.

+ The strawberry mousse and ceps sauce can also be finished cooking one day before and keep in the fridge.

+ The honey tuile can stay crispy kept in an air-tight container for up to 3 days, so make them in advance to save time.

+ Roast the chicken on the day, while the chicken is roasting, prepare the tuna, honey roasted carrot and a little touch for the dessert.

頭盤 | 4 人份

香煎吞拿魚
配柚子西椒莎莎醬

Seared sesame tuna with yuzu and pepper salsa

材料

吞拿魚扒（刺身級別）　200 克
（約 3x3x10 厘米）
黑芝麻　　　　　　　　 50 克
煮食用橄欖油　　　　　 20 毫升
鹽、黑胡椒　　　　　　 適量

柚子西椒莎莎醬

柚子豉油　　　　　　　 50 毫升
初榨橄欖油　　　　　　 30 毫升
新鮮芒果粒　　　　　　 50 克
青西椒（去皮切粒）　　 50 克
紅西椒（去皮切粒）　　 50 克
黑胡椒　　　　　　　　 適量

其他

芝麻菜　　　　　　　　 50 克
沙律生菜　　　　　　　 50 克

做法

柚子西椒莎莎醬

1. 將所有材料放在大碗內拌勻成莎莎醬。

2. 莎莎醬放雪櫃半小時，待用。

完成

3. 用廚房紙印乾吞拿魚扒表面，灑鹽和黑胡椒調味，將黑芝麻均勻鋪上吞拿魚扒表面。

4. 用大火燒熱平底鑊，加入橄欖油，把吞拿魚扒煎至四面金黃。

5. 吞拿魚扒移離平底鑊，立即放在雪櫃裏冷藏。

6. 把已冷藏至冰涼的吞拿魚扒切成方塊，放上莎莎醬和沙律菜享用。

提示

煎吞拿魚扒前，一定要將平底鑊燒至大熱，務求在短時間內將吞拿魚扒煎至金黃。

Seared sesame tuna with yuzu and pepper salsa

INGREDIENTS

- 200 g tuna steak (sashimi grade) (about 3x3x10 cm)
- 50 g black sesame seeds
- 20 ml cooking olive oil
- salt and pepper to judge

YUZU AND PEPPER SALSA

- 50 ml yuzu soya sauce
- 30 ml extra virgin olive oil
- 50 g fresh mango diced
- 50 g green bell pepper, peeled and diced
- 50 g red bell pepper, peeled and diced
- black pepper ground to taste

OTHERS

- 50 g rocket leaves
- 50 g frisée lettuce

METHOD

YUZU AND PEPPER SALSA

1. Combine everything together in a mixing bowl, stir to mix.
2. Keep the salsa in the fridge and let it infuse for half an hour.

TO FINISH

3. For the tuna steak, pat it dry with kitchen paper. Season it with salt and pepper, roll it over the sesame seeds to coat on all sides.
4. Add olive oil in a non-stick pan over high heat, sear the tuna steak until golden brown on all sides.
5. Chill it down in the fridge straight away once removed from the pan.
6. Cut chilled tuna steak into cubes and garnish with the pepper salsa and mixed salad.

COOKING TIPS

+ Make sure the pan is hot enough to sear the tuna until golden brown colour within a short period of time.

主菜 | 4人份

燒法國春雞
伴牛肝菌蜂蜜甘筍

Oven roasted French spring chicken
with honey roasted carrot and ceps sauce

材料

春雞

法國春雞	2 隻
黑松露醬	20 克
黑松露油	10 毫升
新鮮百里香	4 棵
無鹽牛油（軟身）	60 克
鹽、黑胡椒	適量

牛肝菌醬

乾牛肝菌	50 克
乾葱（切碎）	30 克
白蘭地酒	50 毫升
雞湯（看 p.190）	200 毫升
淡忌廉	200 毫升

蜂蜜甘筍

澳洲甘筍（去皮）	150 克
蜂蜜	20 克
新鮮百里香	2 棵
初榨橄欖油	10 毫升
鹽、黑胡椒	適量

其他

小椰菜	4 個

做法

春雞

❶ 手指伸近雞胸和雞皮之間，分開皮肉，把黑松露醬、黑松露油、軟牛油、鹽及黑胡椒均勻搽在雞胸肉上，小心不要弄破雞皮。

❷ 把剩下的醃料抹勻整隻雞，雞腔內塞入新鮮百里香；用棉繩將雞綁好，以固定形態。燒熱平底鑊，下橄欖油，春雞煎至表面金黃。

❸ 焗爐網架放在焗盤上，再將春雞放在網架上。焗爐預熱至攝氏 130-140 度，放入春雞焗 45 分鐘。

❹ 用竹籤插入雞髀位置，如流出的雞汁清澈，即表示春雞已全熟，把已焗熟的春雞靜置 10 分鐘。

牛 肝 菌 醬

⑤ 乾牛肝菌浸在雞湯半小時，過濾，切碎牛肝菌及保留雞湯。

⑥ 用中火將乾葱炒至軟，放入牛肝菌，灒入白蘭地酒，加入雞湯，煮至剩三分之二，加入淡忌廉並煮至濃稠。

蜂 蜜 甘 筍

⑦ 把去皮甘筍切角，與其他材料拌勻。預熱焗爐至攝氏 180 度，把甘筍放在焗盤上焗約 30 分鐘，直至甘筍變軟。

完 成

⑧ 切取春雞雞胸及髀，伴以焓熟的小椰菜葉、蜂蜜甘筍和牛肝菌醬供食。

提示

+ 焗春雞前最少要醃兩小時。

Oven roasted French spring chicken
with honey roasted carrot and ceps sauce

INGREDIENTS

- 2 pcs French spring chicken
- 20 g black truffle paste
- 10 ml black truffle oil
- 4 sprigs fresh thyme
- 60 g unsalted butter, soften
- salt and pepper to judge

CEPS SAUCE

- 50 g dry ceps
- 30 g shallot, chopped
- 50 ml brandy
- 200 ml chicken stock, see p.190
- 200 ml whipping cream

HONEY ROASTED CARROT

- 150 g Australian carrot, peeled
- 20 g honey
- 2 sprigs fresh thyme
- 10 ml extra virgin olive oil
- salt and pepper, to adjust

OTHERS

- 4 pcs brussels sprout

TO PREPARE AND COOK THE CHICKEN

1. Loosen the skin by pushing the fingers between the breast and skin. Push in the truffle paste, oil, softened butter, salt and pepper and gently spread over the breasts without tearing the skin.

2. Rub the rest all over the outside of the chicken and season well. Stuff the fresh thyme in the cavity of the chicken. Tie the chicken with meat string to retain it's good shape. Pan fry it until golden brown all over on a hot pan with olive oil.

3. Then place it on a rack sitting on a baking tray. Roast it in the pre-heated oven (130-140℃) and roast it for 45 minutes.

4. Check the chicken is cooked by piercing the thigh and making sure the juices run clean. Let it rest for about 10 minutes.

CEPS SAUCE

5. Soak the dry ceps in the chicken stock for about half an hour, strain and chop them. Keep the stock.

6. Sauté the chopped shallot in a sauce pan over medium heat until tender, put in ceps, deglaze the pan with brandy. Add the stock and let it reduced by 2/3 and finally add the cream. Let it reduce again until thick.

HONEY ROASTED CARROT

7. Cut the peeled carrot into wedges and mix them with other ingredients. Place them onto a baking tray and roast them for about 30 minutes or until soft in the pre-heated oven (180℃).

TO FINISH

8. Remove the breasts and thighs, serve them with the blanched Brussels sprout leaves, roasted carrots and ceps sauce.

COOKING TIPS

+ Marinate the chicken at least 2 hours before roasting.

燒法國春雞伴牛肝菌蜂蜜甘筍

甜點 | 4 人份

草莓慕絲蜂蜜脆餅

Mille-feuille of strawberry mousse with honey tuile

材料

蜂蜜脆餅

低筋麵粉	25 克
糖霜	25 克
已溶無鹽牛油	25 克
蜂蜜	10 克
雲呢拿豆莢（剖開，刮籽）	半條
蛋白	25 克

草莓慕絲

草莓蓉	100 克
砂糖	15 克
魚膠片（浸凍水）	8 克
檸檬汁	適量
打起淡忌廉	100 毫升

糖漬草莓

新鮮草莓（切粒）	12 顆
薄荷葉（切碎）	5 克
草莓蓉	50 克
糖霜	適量

其他

糖霜（灑面用）	10 克
薄荷葉	少許

做法

蜂蜜脆餅

1. 把低筋麵粉和糖霜一起篩進大碗中，加入溶牛油、蜂蜜及雲呢拿籽拌勻。

2. 加入蛋白，並打發拌勻，放進雪櫃冷藏。

3. 將圓形模板（約 2 毫米厚）放在烘焙墊上，把混合物薄薄的塗在圓形模板內；放入已預熱至攝氏 150 度的焗爐內焗 5-6 分鐘，直至金黃。

草莓慕絲

4. 將一半草莓蓉和砂糖放在小鍋內，加熱至微滾，熄火，加入已浸軟的魚膠片，拌勻。

5. 將混合物轉放入大碗內，並用冰水浸着大碗降溫；加入餘下的草莓蓉和檸檬汁，拌勻。

6. 輕輕捲入已打起的淡忌廉，將混合物倒入鋪上烘焙紙的長方形盤內（長方形盤約 1 厘米厚），放入雪櫃待凝固。

糖漬草莓

⑦ 將所有材料在大碗內拌勻。

完成

⑧ 用圓形餅模切出圓形的草莓慕絲，用三片蜂蜜脆餅如三文治般夾着慕絲。

⑨ 在碟中心放入圓模，舀入糖漬草莓，移離圓模，將草莓慕絲蜂蜜脆餅放在糖漬草莓上，飾上少許糖漬草莓及薄荷葉，灑上糖霜。

 提示

+ 宜在享用甜點前才做「完成」的步驟，因為蜂蜜脆餅如接觸到水氣會很快變軟。

+ 可以將草莓慕絲放到冰格冷藏大約 1 小時，會容易切出圓形。

DESSERT | FOR 4 PORTIONS

Mille-feuille of strawberry mousse with honey tuile

INGREDIENTS

HONEY TUILE

- 25 g cake flour
- 25 g icing sugar
- 25 g unsalted butter, melted
- 10 g honey
- 1/2 pc vanilla pod, split, seeds scraped
- 25 g egg white

STRAWBERRY MOUSSE

- 100 g strawberry puree
- 15 g sugar
- 8 g gelatin leaf, soaked in cold water
- lemon juice to adjust
- 100 ml whipping cream, whipped

STRAWBERRY COMPOTE

- 12 pcs fresh strawberries, diced
- 5 g mint leaves, chopped
- 50 g strawberry puree
- icing sugar to taste

OTHERS

- 10 g icing sugar for dusting
- mint leaves

HONEY TUILE

1. Sieve cake flour and icing sugar together in a mixing bowl. Add the melted butter, honey and vanilla seeds, stir to mix.

2. Finally add the egg white, whisk to combine. Leave it to set in the fridge.

3. Using a round template (about 2 mm thick), spread the mixture thinly on a baking mat and bake them in the pre-heat oven (150℃) for about 5-6 minutes or until they are golden brown.

STRAWBERRY MOUSSE

4. Combine half of the strawberry puree with sugar in a saucepan and bring it to a simmer, remove it from the heat and add the soaked gelatin. Mix well.

5. Transfer the mixture into a mixing bowl sitting over ice water, add the remaining half of the strawberry puree, lemon juice and stir well.

6. Finally fold in the whipped cream, pour the mixture onto a one-cm-thick rectangular tray lined with parchment paper. Let it set in the fridge.

STRAWBERRY COMPOTE

7. Mix everything together in a mixing bowl.

TO FINISH

8. Cut out the mousse by using a round cutter and sandwich them by using the honey tuile.

9. Mould the strawberry compote at the center of a plate and transfer the strawberry sandwich onto the compote. Finally garnish with the mint leaves and strawberry compote. Dust it with icing sugar.

COOKING TIPS

+ It would be better to assemble the dessert at the last minute as the honey tuiles will go soggy very quick once they have contacted with liquid.

+ Freeze the strawberry mousse for about one hour for easier cutting.

草莓慕絲蜂蜜脆餅

SET/02

spring menu

香 煎 帶 子
伴 帶 子 香 腸

Scallops in 2 ways

構 思 　HOW IS IT DESIGNED ?

我是海鮮的超級擁躉，尤其是甲殼類，如帶子和蝦，兩者都是春天的時令食材，味道清鮮，烹調時間宜短，適宜配合一些味道較淡和質感較嫩的食材，如豆、番茄、甘筍和椰菜花等。白焗、快煎、慢煮都能夠保留海鮮的養分和味道，這些烹調法我亦應用在這餐單上。

相對於清鮮的頭盤和主菜，我用焦糖榛子芭菲配焦糖啤梨作甜點結尾。

I am a big fan of seafood especially the shellfish like scallops and prawns which are both available during the spring time. They are mild in flavour, so it is better to cook less and serve with something subtle and delicate like spring baby vegetables such as peas, tomatoes, carrots and cauliflowers. Poaching, searing and slow-cooking, which can retain the most nutrients and flavour from the seafood, are applied into this menu. Pear and hazelnut are the perfect match to round up the lightless of the starter and main course.

Main Dish

主菜

Dessert

甜點

番 茄 春 蔬
伴 大 蝦

Pot of poached prawns in olive oil broth with
tomatoes and spring vegetables

焦 糖 榛 子 芭 菲
配 焦 糖 啤 梨

Hazelnut praline parfait
with caramelized pear

預 先 準 備 TIPS FOR SMART PREPARATION

+ 椰菜花蓉可以早一天準備，放入雪櫃冷
  藏保鮮。

+ 帶子香腸可以在進餐當日預早捲起冷
  藏，享用前才煮熟。

+ 芭菲亦可以預早製作，在冰格裏雪藏。

+ 酥脆餅皮可以預早焗好，放在密封瓶內
  可保鮮 1 星期。

+ The cauliflower puree can be made one
  day in advance and keep in the fridge.

+ The scallop sausage can also be rolled
  up before and keep in the fridge for
  cooking during the day of service.

+ The parfait can definitely be prepared in
  advance and keep in the freezer.

+ The filo leaves can also be baked
  beforehand and keep in an air tight
  container for up to a week.

頭盤 | 4 人份

香煎帶子
伴帶子香腸

Scallop in 2 ways

 材料

帶子（刺身級別）	4 隻
蝦湯（ 看 p.192）	100 毫升
無鹽牛油（雪硬）	30 克

椰菜花蓉

椰菜花（切碎）	100 克
牛奶	200 毫升
無鹽牛油	20 克
檸檬汁	適量
鹽、黑胡椒	適量

帶子香腸

帶子肉（刺身級別）（雪硬）	120 克
淡忌廉	12 毫升
鹽、黑胡椒	適量
牛油	10 克

做法

椰菜花蓉

① 椰菜花、牛奶、鹽、黑胡椒放入鍋內，拌勻，煮約 15 分鐘直至椰菜花變軟。隔去牛奶，牛奶留作打起泡沫之用；椰菜花加牛油和檸檬汁放在攪拌機內打成椰菜花蓉。

帶子香腸

② 把帶子肉放在食物處理器內打成滑漿，下淡忌廉和調味料拌勻。

③ 用保鮮紙捲起帶子漿成香腸形狀。

提示

＋ 煎帶子前要以廚房紙抹乾水分，並要確保平底鑊非常熱，才可把帶子煎成金黃色。

煮帶子香腸

④ 外鍋加水，內鍋放在外鍋內，內鍋水分維持在攝氏 60 度，把帶子香腸放在內鍋慢煮 15 分鐘。

⑤ 移去保鮮紙，用廚房紙印乾帶子香腸，之後用牛油煎香。

完成

⑥ 帶子調味，在平底鑊煎至兩面金黃。

⑦ 將帶子移離平底鑊，加入蝦湯，煮剩 2/3，之後加入凍牛油煮成牛油蝦汁。

⑧ 加熱煮椰菜花的牛奶，用手動攪拌器打至起泡。

⑨ 椰菜花蓉舀進碟內，放入煎帶子及帶子香腸，配牛油蝦汁和泡沫享用。

Scallop in 2 ways

INGREDIENTS

- 4 pcs scallop, sashimi grade
- 100 ml shrimp stock, see p.192
- 30 g unsalted butter, chilled

CAULIFLOWER PUREE

- 100 g cauliflower, roughly chopped
- 200 ml milk
- 20 g unsalted butter
- lemon juice to taste
- salt and pepper to taste

SCALLOP MOUSSE SAUSAGE

- 120 g scallop meat, sashimi grade, well chilled
- 12 ml whipping cream
- salt and pepper to taste
- 10 g butter

COOKING TIPS

+ In order to get a golden brown colour of the scallops, make sure they are patted dry enough, and the frying pan is super hot.

METHOD

CAULIFLOWER PUREE

1. In a sauce pan, mix cauliflower, milk, salt and pepper together. Cook it for about 15 minutes or until cauliflower is soft. Strain the cauliflower and blend it together with butter and lemon juice into a puree. Reserve the cooking liquid for making foam.

SCALLOP MOUSSE SAUSAGE

2. Put the scallop meat in a processor and blend until smooth, add the cream and seasoning, mix well.

3. Roll the mousse into sausages by using cling film.

TO COOK

4. Slow-cook the scallop mousse in the bain marie of 60℃ for about 15 minutes.

5. Remove the cling film and pat dry with kitchen paper, then pan fry with butter until brown.

TO FINISH

6. Season the scallops and pan fry until golden brown on both sides.

7. Remove the scallops and add the shrimp stock, let it reduce by 2/3. Add the chilled butter to finish the sauce.

8. Warm up the cauliflower cooking liquid, use a hand blender to make foam.

9. Serve the pan-fried scallops and sausage with the shrimp butter sauce and the foam.

香煎帶子伴帶子香腸

主菜│4 人份

番茄春蔬伴大蝦
Pot of poached prawns
in olive oil broth
with tomatoes and spring vegetables

材料

		橄欖油蝦湯	
新鮮大蝦	400 克	蒜片	10 克
車厘茄（切半）	100 克	初榨橄欖油	30 毫升
迷你甘筍（去皮）	8 條	蝦湯（看 p.192）	10 克
珍珠洋葱（去衣）	20 粒	番紅花	0.5 克
綠、黃色意大利青瓜	200 克	檸檬皮	1/4 個
羊肚菌	4 個	法國芫茜（切碎）	10 克
粟米芯	8 條	鹽、黑胡椒	適量
迷你蘆筍	8 條		
本菇	8 朵		

做法

橄欖油蝦湯

① 湯鍋內下橄欖油，用小火煎蒜片至稍黃，下蝦湯、番紅花及檸檬皮，繼續煮約 10 分鐘，下鹽和黑胡椒調味。

預備蝦和蔬菜

② 蝦去殼、去腸，殼可留作煮湯用。

③ 把蔬菜切成所需的形狀。

完成

④ 每樣蔬菜分別在蝦湯中煮熟，保暖備用，下蝦煮至剛熟。

⑤ 蝦和蔬菜上碟，倒入熱橄欖油蝦湯，灑上法國芫茜碎及初榨橄欖油。

提示

+ 留蝦殼用作煮蝦湯用。

+ 蝦肉不要煮太久，否則肉會變硬和韌。

Pot of poached prawns in olive oil broth with tomatoes and spring vegetables

INGREDIENTS

- 400 g fresh prawns, heads and shells removed
- 100 g cherry tomatoes, cut into halves
- 8 pcs baby carrot, peeled
- 20 pcs pearl onions, peeled
- 200 g green and yellow zucchini
- 4 pcs morel mushrooms
- 8 pcs baby corn
- 8 pcs baby asparagus
- 8 pcs shimeji mushrooms

OLIVE OIL SHRIMP BROTH

- 10 g garlic slices
- 30 ml extra virgin olive oil
- 10 g shrimp stock, see p.192
- 0.5 g saffron threads
- 1/4 pc lemon zest
- 10 g chervil, chopped
- salt and pepper to taste

METHOD

OLIVE OIL BROTH

1. Cook the garlic slices with olive oil in a pot over low heat until lightly brown, add the shrimp stock, saffron threads and lemon zest, let it simmer for about 10 minutes, season with salt and pepper.

TO PREPARE THE PRAWNS AND VEGETABLES

2. Remove the shells and de-vein from the prawns. Keep the shells to make the stock.

3. Cut the vegetables into the required sizes and shapes.

TO FINISH

4. Cook the vegetables individually in the broth and keep warm aside. Poach the prawns until just cooked.

5. Arrange the prawns and vegetables on a plate and pour the hot broth over, sprinkle the chopped chervil on top and drizzle with extra virgin olive oil.

COOKING TIPS

+ Keep the prawns shells for making the stock.

+ Don't overcook the prawns in the broth, otherwise they will be tough and chewy.

番茄春蔬伴大蝦

甜點 | 4 人份

焦糖榛子芭菲
配焦糖啤梨

Hazelnut praline parfait
with caramelized pear

材料

焦糖榛子芭菲

焗香榛子	30 克
幼砂糖（1）	60 克
幼砂糖（2）	80 克
水	20 毫升
蛋白	100 克
打起淡忌廉	200 毫升

焦糖啤梨

幼砂糖	200 克
啤梨（去皮去芯切半）	4 個

酥皮

菲路餅皮（filo pastry）	6 塊
無鹽牛油（已溶）	適量
糖霜（灑面用）	適量

裝飾

法國芫茜	適量

做法

焦糖榛子芭菲

1. 將幼砂糖（1）灑在熱鍋上，煮成焦糖，下榛子，拌勻。焦糖榛子放在墊上冷卻。

2. 焦糖榛子冷卻後，切成細件。

3. 在小鍋中下幼砂糖（2）和水加熱煮滾至攝氏 110 度。

4. 將熱糖漿慢慢倒入已打起的蛋內，不斷攪拌直至完全融合，繼續攪拌 3 分鐘直至冷卻。

5. 將焦糖榛子及打起淡忌廉倒入蛋白內，攪勻，並均勻鋪在已放烘焙紙的焗盤上，放入冰格冷藏，雪硬後切成需要的形狀。

焦糖啤梨

6. 用中火加熱小鍋，慢慢灑入幼砂糖，並持續移動小鍋至幼砂糖溶化。

7. 將焦糖煮成琥珀色，下啤梨煮至剛剛變軟，但仍能維持形狀。

8. 把焦糖啤梨放在盤上冷卻至室溫，剩下的焦糖漿留用。

酥皮

⑨ 大焗盤上噴油，放一塊菲路餅皮，在餅皮塗上已溶牛油，灑上糖霜；放下另一塊菲路餅皮，塗上已溶牛油，灑上糖霜，放入雪櫃裏冷藏定形。

⑩ 把餅皮切成需要形狀，放入已預熱至攝氏 180 度焗爐焗至金黃，取出待涼。

完成

⑪ 用三片酥皮如三文治般夾着焦糖榛子芭菲。

⑫ 放上焦糖啤梨，旁邊放焦糖榛子芭菲。

⑬ 如有需要，可下焦糖糖漿，最後以法國芫茜裝飾。

提示

+ 如果不介意吃生蛋白，可以只用蛋白和糖打起作芭菲。

+ 蛋白混合物必須冷卻後才與忌廉混合，否則會很容易塌下。

Hazelnut praline parfait with caramelized pear

INGREDIENTS

HAZELNUT PRALINE PARFAIT

- 30 g hazelnuts, toasted
- 60 g sugar (1)
- 80 g sugar (2)
- 20 ml water
- 100 g egg white
- 200 ml whipping cream, whipped

CARAMELIZED PEARS

- 200 g castor sugar
- 4 pcs pear, peeled, cored and halved

FILO LEAVES

- 6 sheets filo pastry
- unsalted butter, melted, to judge
- icing sugar for dusting, to judge

GARNISHES

- a few sprigs chervil

HAZELNUT AND PRALINE PARFAIT

1. Heat a pan on the stove, then sprinkle the sugar (1) on and let it caramelize. Add the toasted hazelnuts and stir together, then pour the caramelized nuts onto a mat and allow to cool.

2. When the praline is cold, chop it into small pieces.

3. In a sauce pan, combine sugar (2) and water and bring them to a boil until it reaches to about 110℃.

4. Then slowly and steadily pour the hot syrup into the soft whipped egg white. Keep whisking until all the syrup has been incorporated into the egg white. Continue whisking for a further 3 minutes until cool.

5. Fold the chopped praline and whipped cream into the meringue mixture. Spread into a tray lined with baking paper and freeze. Cut it into the required shape when frozen.

CARAMELIZED PEARS

6. Place a sauce pan over medium heat, gradually sprinkle on sugar, shake the pan to let it melt.

7. Cook the caramel to a golden amber colour, then add the pears and continue to cook until they are just tender but firm and holding their shape.

8. Transfer the pears to a tray and allow cooling to room temperature, and reserving the caramel for saucing.

FILO LEAVES

9. Lay a sheet of filo pastry onto a large baking tray coated with oil spray. Brush the pasty with melted butter, then dust with icing sugar. Place another sheet on top and repeat the process. Chill it in the fridge to set.

10. Cut them into the required shapes and baking them in the pre-heated oven (180℃) until golden brown. Allow to cool.

TO FINISH

11. Sandwich hazelnut and praline parfait by using the filo leaves.

12. Transfer the pears onto the plate, arrange crispy hazelnut and praline parfait aside.

13. Drizzle with extra caramel sauce if needed. Decorate with chervil.

COOKING TIPS

+ The egg white can be simply whipped up with sugar (2) while making the parfait if you don't really care about eating raw egg white.

+ The egg white mixture must be cooled down before fold in the whipped cream otherwise it will collapse easily.

焦糖榛子芭菲配焦糖啤梨

SET/03

spring
menu

法 式 豬 肉 雞 肝 醬 配 茴 香 沙 律

Chicken liver and pork terrine
with fennel salad

構 思   HOW IS IT DESIGNED ?

如你比較喜歡吃肉,這個套餐是春日的好選擇。建議以 fruity 和 full body 的紅酒配搭頭盤;主菜,我採用巴馬火腿捲起味道清甜而健康的雞胸,煎熟享用;最後,我用甜美的朱古力撻,配上略帶酸的紅莓雪葩,圓滿而滿足。

If you don't like seafood, but something meaty, this menu is a good choice during the spring time. A glass of fruity and full body red wine is recommended to start with the cold terrine which is complimented with a light and aromatic fennel salad. The chicken breast here is, healthy and light in flavour, simply wrapped with ham and pan-fried to cook. To round up this menu, it is nothing better to go with something sweet and sour like chocolate tart with raspberry sorbet.

Main Dish

主菜

巴 馬 火 腿 雞 胸 卷
伴 芥 末 薯 仔

Chicken breast roll wrapped with Parma ham
served with potato remoulade

Dessert

甜點

朱 古 力 撻
配 紅 莓 雪 葩

Chocolate tart
with raspberry sorbet

預 先 準 備 TIPS FOR SMART PREPARATION

+ 豬肉雞肝醬需要徹底煮熟並冷藏過夜。

+ 可以預早兩個小時捲起巴馬火腿雞胸卷，放入雪櫃冷藏。

+ 可以預先做好撻皮並焗熟；雪葩可以在享用前數天做好，並放入冰格冷藏。

+ The chicken liver and pork terrine has to be fully cooked and keep it in the fridge to set overnight.

+ The chicken breast roll can be wrapped two hours in advance before service and let it set in the fridge.

+ The tart case can be made and baked blinded, and finish the sorbet and keep in the freezer a few day in advance.

頭盤 | 8人份

法式豬肉雞肝醬
配茴香沙律

Chicken liver and pork terrine
with fennel salad

材料

豬肉雞肝醬

新鮮雞肝（打成漿，過篩）	350 克
免治豬肉	350 克
生香腸肉	350 克
意大利芫茜（切碎）（或免）	5 克
細香葱（切碎）	5 克
薑粉	2 克
肉桂粉	2 克
開心果	30 克
白蘭地酒	20 毫升
雪利酒	20 毫升
煙肉／意大利鹽醃腩肉	150 克
鹽、黑胡椒	共 3 克

其他

已切好鵝肝（煎好，冷卻）	3 塊

沙律

芝麻菜（已清洗）	200 克
茴香頭（切薄片）	2 個
初榨橄欖油	50 毫升
意大利黑醋	30 毫升
鹽、黑胡椒	適量

做法

豬肉雞肝醬

1. 將所有材料（除煙肉）放在大碗內，拌勻成醬狀。

2. 取一個 12 x 3.5 x 4 吋的焗盆，鋪上煙肉，倒入一半豬肉雞肝醬，放上已煎好的鵝肝，加入餘下的豬肉雞肝醬。

3. 將煙肉覆在醬面，鋪上烘焙紙，蓋上焗盆蓋，用水浴法，以

攝氏 160 度焗 50 至 60 分鐘，直至所有材料熟透，內部的溫度大約為攝氏 78 度。

4. 從焗爐取出焗盆，倒走焗盆內過多的汁液；更換另一張烘焙紙，用重物壓着已熟透的豬肉雞肝醬，待冷卻，宜放在雪櫃內冷藏一夜。

提示

十 豬肉雞肝醬需要冷藏過夜才切片，否則會很容易弄散。

沙律

5. 大碗內下橄欖油及意大利黑醋，拌勻，下鹽及黑胡椒調味。

6. 將沙律汁灑在沙律菜上，留少許備用。

完成

7. 將豬肉雞肝醬從焗盆倒出，切片，上碟，以沙律菜伴碟，灑上沙律汁享用。

Chicken liver and pork terrine with fennel salad

INGREDIENTS

MEAT LOAF

- 350 g fresh chicken liver, puree and sieved
- 350 g pork, minced
- 350 g raw sausage meat
- 5 g Italian parsley (optional), chopped
- 5 g chive, chopped
- 2 g ginger powder
- 2 g cinnamon powder
- 30 g pistachio
- 20 ml brandy
- 20 ml sherry wine
- 150 g smoked bacon/pancetta slices
- 3 g salt and pepper

OTHERS

- 3 pcs pre-cut foie gras, pan fried and cool down

SALAD

- 200 g rocket leaves, cleaned
- 2 pcs fennel head, thinly sliced
- 50 ml extra virgin olive oil
- 30 ml balsamic vinegar
- salt and pepper to taste

COOKING TIPS

+ The terrine must be chilled overnight before slicing, otherwise it would break easily.

METHOD

MEAT LOAF

1. Combine everything together (except the bacon) in a mixing bowl, and mix well.

2. Line a 12 x 3.5 x 4 inch loaf pan with bacon strips and pack in half of the mixture, then arrange the pan-fried foie gras in the middle and cover with the remaining mixture.

3. Close the bacon strips on top. Cover the pan with parchment paper and place it in a water bath. Bake at 160°C for about 50 to 60 minutes or until it is cooked inside (the temperature should be around 78°C).

4. Remove from heat. Pour off excessive juice, cover with a new parchment paper. Place a weight on top while cooling. Best to cool overnight in the fridge.

SALAD

5. To make the dressing, combine olive oil and balsamic vinegar in a mixing bowl. Whisk to combine, season with salt and pepper.

6. Add the dressing to the salad leaves, keep some for drizzling during plating.

TO SERVE

7. Remove the terrine from the mould and cut it into slices, arrange them on a plate. Put the salad on the side. Drizzle with extra dressing around.

法式豬肉雞肝醬配茴香沙律

41

主菜 | 4 人份

巴馬火腿雞胸卷
伴芥末薯仔

Chicken breast roll wrapped with Parma ham
served with potato remoulade

雞 卷

法國春雞胸	2 件
巴馬火腿或意大利火腿	4 片
新鮮羅勒葉	6 片
鹽、黑胡椒	適量
橄欖油	10 毫升

辣 蛋 黃 醬 新 薯

新薯（洗淨）	6 個
青瓜（切粒）	1/3 個
西芹（切粒）	半條
根芹（或免）（切粒）	50 克
法式芥末醬	10 克
酸忌廉	30 克
酸青瓜（切粒）	50 克
水瓜鈕（切碎）	10 克
番茜或他拉根香草（切碎）	5 克
白酒醋	15 毫升
初榨橄欖油	15 毫升
鹽、黑胡椒	適量

羅 勒 白 醬

乾葱（切碎）	50 克
白酒	50 毫升
雞湯	150 毫升
淡忌廉	50 毫升
新鮮羅勒（切碎）	10 克

裝 飾

豌豆苗	80 克

 做法

準 備 雞 卷

1 雞胸去皮，以鹽及黑胡椒調味。工作枱上鋪保鮮紙，在保鮮紙上放兩片巴馬火腿及羅勒葉，再放上雞胸，捲實成雞卷，放在雪櫃冷藏固定雞卷的形狀。

辣 蛋 黃 醬 新 薯

2 薯仔連皮隔水蒸 20 分鐘，或至變軟，去皮切粒。

3 大碗內放入其餘材料，拌勻，然後與薯粒一起拌勻，保溫。

巴
馬
火
腿
雞
胸
卷
伴
芥
末
薯
仔

43

羅勒白醬

④ 用中火將乾蔥碎炒至變軟，灒下白酒，加入上湯，煮至剩約1/3，下淡忌廉，煮至略呈濃稠。

⑤ 下羅勒葉拌勻，保溫。

完成

⑥ 拆開雞卷的保鮮紙；易潔平底鑊內加入橄欖油，用中火將雞卷煎至表面金黃。

⑦ 將雞卷放在焗盤上，放入已預熱攝氏 120 度的焗爐內焗 15 至 20 分鐘，或至熟透。

⑧ 將雞卷從焗爐取出，靜置雞卷 5 至 10 分鐘，切件，伴辣蛋黃醬新薯及羅勒白醬，飾上豌豆苗。

提示

＋ 以低溫（約攝氏 120 度）焗雞卷，可以保持雞肉的汁液和嫩滑的口感。

Chicken breast roll wrapped with Parma ham served with potato remoulade

INGREDIENTS

CHICKEN BREAST ROLL

- 2 pcs French spring chicken breast
- 4 slices Parma ham/prosciutto
- 6 fresh basil leaves
- salt and pepper to judge
- 10 ml cooking olive oil

POTATO REMOULADE

- 6 pcs new potatoes, cleaned
- 1/3 pc cucumber, diced
- 1/2 stick celery, diced
- 50 g celeriac (optional), diced
- 10 g Dijon mustard
- 30 g sour cream
- 50 g cornichons, diced
- 10 g capers, chopped
- 5 g parsley or tarragon, chopped
- 15 ml white wine vinegar
- 15 ml extra virgin olive oil
- salt and pepper to taste

BASIL CREAM

- 50 g shallot, chopped
- 50 ml white wine
- 150 ml chicken stock
- 50 ml whipping cream
- 10 g fresh basil, chopped

GARNISHES

- 80 g pea sprout

TO PREPARE THE CHICKEN BREAST ROLL

1. Remove the skin from the breast and season with salt and pepper. Lay 2 slices of Parma ham and basil leaves on the cling film sticking on a table. Arrange the breast in the middle and roll it up into a sausage. Keep chilled to set.

POTATO REMOULADE

2. Steam the potatoes with the skin on for about 20 minutes or until they are tender. Peel the skin off and cut them into dices.

3. Mix all the remaining ingredients in a mixing bowl, stir to mix. Add this sauce into the diced potatoes, mix well, keep warm.

BASIL CREAM

4. Sweat the chopped shallot in a sauce pan over medium heat until tender, deglaze the pan with white wine. Add the stock and let it reduced by 2/3 and add the cream. Let it reduce again until lightly thick.

5. Finally add the chopped basil and keep warm.

TO FINISH

6. Remove the cling film from the chicken breast rolls and pan fry it with olive oil in a non-stick pan over medium heat until golden brown all around.

7. Transfer the rolls onto a baking tray and put it into the pre-heated oven 120°C for about 15 to 20 minutes or until cooked.

8. Let the chicken rest for about 5-10 minutes and cut it into slices, serve them with the potato remoulade and sauce. Garnish with pea sprout.

COOKING TIPS

+ Cook the chicken breast at a low temperature (120°C) to retain its moisture and tenderness.

巴馬火腿雞胸卷伴芥末薯仔

45

甜點 | 4 人份（24 x 10 x 2.5 厘米撻模）

朱古力撻配紅莓雪葩
Chocolate tart with raspberry sorbet

 材料

杏仁撻皮

糖霜	27 克
無鹽牛油（切粒）	36 克
杏仁粉	9 克
低筋麵粉	66 克
蛋漿	15 克
雲尼拿油	0.5 克
鹽	0.5 克

朱古力醬

黑朱古力（72% 或更高）（已溶）	144 克
淡忌廉	120 毫升
軟無鹽牛油	22 克
蜂蜜	6 克

紅莓雪葩

水	80 毫升
砂糖	40 克
葡萄糖漿	40 克
急凍或新鮮紅莓	200 克
檸檬汁	適量調味

裝飾

薄荷葉	數片
開心果（打碎）	5 克
新鮮紅莓	30 粒

做法

杏仁撻皮 | 紅莓雪葩

杏仁撻皮

1. 將糖霜、杏仁粉、低筋麵粉和鹽倒入大碗內，拌勻，加入牛油粒，用手揉或用有鈎的打蛋器打至如麵包糠般。

2. 下蛋漿及雲尼拿油，攪拌成為軟麵糰（不要過分搓揉麵糰），放在雪櫃冷藏。

3. 取出麵糰，擀至 2-3 毫米厚，鋪在焗盤內，放入雪櫃冷藏；冷藏後，在撻皮面放上烘焙紙和耐熱珠，放入已預熱攝氏 180 度焗爐焗 10 分鐘。

4. 把焗爐調校至 150 度焗 20 分鐘，移去烘焙紙和耐熱珠，焗至金黃色。

朱古力醬

5. 坐暖淡忌廉，分三次倒入已溶黑朱古力，攪拌勻後再拌入軟牛油及蜂蜜。

紅莓雪葩

6. 把水、砂糖、紅莓放在鍋中煮滾，再煮 5 分鐘後移離火爐，倒入葡萄糖漿和檸檬汁，拌勻。

7. 用打蛋器快速攪打，過幼篩，並用冰水浴冷卻。

8. 將混合物倒進雪糕機內攪約 25 至 30 分鐘。

完成

9. 朱古力醬倒入撻皮上，放上新鮮紅莓。

10. 朱古力撻放在雪櫃冷藏，切件上碟，伴紅莓雪葩及裝飾。

提示

+ 杏仁撻皮可以預早幾天做好放入雪櫃冷藏，或把撻皮焗熟後放在室溫密封容器內。

Chocolate tart with raspberry sorbet

INGREDIENTS

ALMOND SWEET PASTE

- 27 g icing sugar
- 36 g unsalted butter, cut into cubes
- 9 g almond powder
- 66 g cake flour
- 15 g whole eggs
- 0.5 g vanilla extract
- 0.5 g salt

CHOCOLATE GANACHE FILLING

- 144 g dark chocolate 72% or above (melted)
- 120 ml whipping cream
- 22 g unsalted butter, soften
- 6 g honey

RASPBERRY SORBET

- 80 ml water
- 40 g sugar
- 40 g glucose
- 200 g frozen/fresh raspberry
- lemon juice to taste

GARNISHES

- a few pcs mint leaf
- 5 g pistachio, grated
- 30 pcs fresh raspberry

COOKING TIPS

+ The sweet dough for making the tart case can be made a few days in advance, keep it in the fridge for later use. Or the tart can be finished baking and kept in an air-tight container at room temperature.

METHOD

ALMOND SWEET PASTE

1. Combine icing sugar, almond powder, cake flour and salt in a mixing bowl, add the butter cubes and rub them with hand or hand mixer fitted with hook until dry and crumbly.
2. Add the eggs and vanilla extract to the mixture, mix it thoroughly to form soft dough. Remember don't knead it too much. Chill it down in the fridge.
3. Take the dough out from the fridge and roll it out to about 2-3 mm thick, transfer it to a baking tin and line it on properly. Chill it down in the fridge again and bake it blind in the pre-heated oven (180°C) for about 10 minutes.
4. Turn down the temperature to around 150°C, remove the baking paper and the stones, bake another 20 minutes until it is golden brown.

CHOCOLATE GANACHE FILLING

5. Warm the cream and pour it into the melted chocolate in 3 times, stir constantly, finally add the butter and honey.

RASPBERRY SORBET

6. Combine water, sugar and raspberry in a sauce pan and bring it to a boil. Cook it for about 5 minutes and remove it from the heat. Stir in glucose and lemon juice.
7. Blitz it with a hand blender and pass it through a fine sieve. Cool it down over ice water.
8. Pour the mixture into ice-cream machine and churn it for about 25 to 30 minutes.

TO FINISH

9. Pour the chocolate ganache at the bottom of the tart case, and then arrange the fresh raspberries evenly on top.
10. Chill it in the fridge until set. Cut it into portions and serve them with the sorbet and the garnishes.

朱古力槤配紅莓雪葩

49

SET/04

spring menu

洋 葱 蘑 菇 撻
伴 芝 麻 菜 沙 律

Mushroom and onion quiche
with rocket salad

構 思 HOW IS IT DESIGNED ?

如果你不是全素食者，這個蛋奶素的套餐絕對適合你在春天享用。

我用酥脆美味的洋葱蘑菇撻伴以芝麻菜沙律，來為這個春日素餐展開序幕；再配以鬆軟多汁、味道淳樸的法式焗雜菜，絕對為這素食增添不少色彩。味道濃郁的豆漿米布甸，配以清新的芒果和香茅冰沙，讓味道平衡得宜。

If you are not a vegan, this vegetarian menu definitely suits you during the spring time. Flaky and aromatic vegetables quiche with peppery rocket salad kicks off the menu, after that, something moist, natural and clean flavour from Ratatouille replenish the meal. Rice pudding itself is quite rich; on the other hand, the lemongrass granita and fresh mango can cut off the richness to round up the menu.

Main Dish

主菜

法 式 焗 雜 菜

Ratatouille

Dessert

甜點

豆 漿 米 布 甸
配 香 茅 冰 沙

Soya milk rice pudding
with lemongrass granita

預 先 準 備 TIPS FOR SMART PREPARATION

+ 洋葱磨菇撻的撻皮可以預早一天焗熟。

+ 豆漿米布甸也可以預早一天弄好,倒進玻璃杯冷藏。

+ The quiche tart case can be prepared and baked
  blinded one day in advance.

+ The rice pudding can be prepared one day before and
  kept in the serving glasses in the fridge.

頭盤 | 4 人份（13 x 4 x 2 厘米撻模）

洋葱蘑菇撻
伴芝麻菜沙律

Mushroom and onion quiche
with rocket salad

材料

撻皮

低筋麵粉	90 克
鹽	1.5 克
砂糖	6 克
即磨黑胡椒	少許
無鹽牛油（雪硬）	54 克
蛋漿	18 克

餡料

白洋葱（切片）	400 克
橄欖油	10 毫升
新鮮百里香	2 棵
蒜茸	20 克
白蘑菇	100 克
淡忌廉	110 毫升
蛋漿	90 克
巴馬臣芝士	15 克
鹽、黑胡椒	適量

其他

沙律生菜葉	40 克
芝麻菜	80 克
意大利黑醋	20 毫升
初榨橄欖油	40 毫升
鹽、黑胡椒	適量

做法

撻皮

1. 低筋麵粉篩進大碗內，加入砂糖、鹽和黑胡椒拌勻，下牛油並用手揉成麵包糠狀。

2. 下蛋漿並拌勻成軟麵糰。

3. 用保鮮紙包起，放雪櫃冷藏 20 分鐘。

4. 將粉糰壓平，再放在撻模內壓好，放雪櫃冷藏大約 10 分鐘。面放烘焙紙和焗珠，以攝氏 180 度焗 10 分鐘，調校至 150 度再焗 20 分鐘，直至撻皮微黃。

餡料

5. 平底鑊加熱下橄欖油和百里香,用小火炒洋蔥片約 30 分鐘,直至洋蔥金黃。

6. 用厚底平鑊、中大火炒香蒜茸和蘑菇,下鹽和黑胡椒粉調味,炒約 8 分鐘直至蘑菇微黃,上碟待涼。

7. 蛋漿、淡忌廉、鹽及黑胡椒倒入大碗內,攪勻。

洋蔥蘑菇撻

8. 將蘑菇和焦糖洋蔥鋪在撻皮上,倒入忌廉蛋漿,灑上芝士。

9. 以攝氏 150 度焗洋蔥蘑菇撻,直至脹起及金黃色,約需時 20 分鐘。

完成

10. 將沙律菜、橄欖油和意大利黑醋拌勻,伴暖暖的洋蔥蘑菇撻,並灑下黑醋和橄欖油享用。

提示

+ 建議先將撻皮焗好,才加入餡料再焗,這可以保持撻皮鬆脆。

Mushroom and onion quiche with rocket salad

INGREDIENTS

FILLING

- 400 g white onions, sliced
- 10 ml olive oil
- 2 sprigs fresh thyme
- 20 g garlic, crushed
- 100 g white button mushrooms
- 110 ml whipping cream
- 90 g whole eggs
- 15 g parmesan cheese to taste
- salt and pepper to taste

SAVOURY QUICHE PASTE

- 90 g cake flour
- 1.5 g salt
- 6 g sugar
- pinch black pepper, grind
- 54 g unsalted butter, chilled
- 18 g whole egg

OTHERS

- 40 g frisée salad leaves
- 80 g rocket leaves
- 20 ml balsamic vinegar
- 40 ml extra virgin olive oil
- salt and pepper to taste

COOKING TIPS

+ To retain the crispiness of the tart case, it would be much better to bake the tart case blinded before pour in the filling for the second time baking.

METHOD

TO MAKE THE TART

1. Combine the sieved flour, sugar, salt and pepper together in a mixing bowl. Add the chilled butter cubes and rub them to achieve a breadcrumb texture.
2. Then add the eggs, carefully mix into a smooth paste.
3. Wrap it with cling film and keep it in the fridge for about 20 minutes.
4. Roll the paste out and line onto the tart moulds. Bake them blind at 180°C for 10 minutes; turn it down to 150°C for another 20 minutes or until brown.

TO MAKE THE FILLING

5. Cook the sliced onions with olive oil and fresh thyme in a pan over low heat until golden brown, it takes around 30 minutes.
6. Sauté the mushrooms in heavy large skillet over medium-high heat, season with salt and pepper, cook until lightly brown, about 8 minutes. Transfer to a plate; spread out to cool slightly.
7. Whisk eggs, whipping cream, salt and pepper in a large bowl to blend.

TO FINISH

8. Arrange the sautéed mushrooms and caramelized onions on the cooked tart, then pour the filling in. Sprinkle the cheese on.
9. Bake the quiche (150°C) until puffed, golden brown, and just set in center, about 20 minutes.

TO SERVE

10. Mix the salad leaves with balsamic vinegar and olive oil, serve it with the warm quiche. Drizzle balsamic vinegar and olive oil dressing around.

洋葱蘑菇撻伴芝麻菜沙律

55

🌿 主菜 | 4 人份

法式焗雜菜
Ratatouille

（材料）

番茄醬

新鮮番茄	800 克（去皮切粒）
洋蔥	120 克（切碎）
蒜肉	2 粒（切碎）
初榨橄欖油	40 毫升
茄膏	40 克
罐頭番茄	200 克
羅勒	20 克
鹽、黑胡椒	適量

雜菜

茄子	1 條（切成半厘米厚）
綠色意大利青瓜	1 條（切成半厘米厚）
黃色意大利青瓜	1 條（切成半厘米厚）
番茄	4 個（切成半厘米厚）
檸檬皮	1/4 個
初榨橄欖油	5 克
鹽、黑胡椒	適量

（提示）

雜菜不宜焗得太久，否則會失去光澤和豐富的顏色。

做法

番茄醬

1. 用中火燒熱平底鑊，下橄欖油和蒜粒，將蒜粒炒至微黃，下洋葱炒至變軟，加入新鮮番茄並煮至差不多乾水。

2. 下茄膏、罐頭番茄及羅勒，煮10分鐘，下鹽和黑胡椒調味，備用。

燴雜菜

3. 將番茄醬舀入容器內，抹平，相間地排入雜菜，下鹽和黑胡椒調味。

4. 放入已預熱攝氏180度之焗爐，焗15至20分鐘，直至所有蔬菜變軟。

上菜

5. 刨下檸檬皮茸，澆橄欖油，即可享用。

Ratatouille

INGREDIENTS

TOMATO SAUCE

- 800 g fresh tomatoes, peeled and diced
- 120 g onion, chopped
- 2 cloves garlic, chopped
- 40 ml extra virgin olive oil
- 40 g tomato paste
- 200 g tinned tomato
- 20 g basil, chopped
- salt and pepper to taste

MIXED VEGETABLES

- 1 pc eggplant, cut into 1/2 cm thick
- 1 pc green zucchini, cut into 1/2 cm thick
- 1 pc yellow zucchini, cut into 1/2 cm thick
- 4 pcs tomatoes, cut into 1/2 cm thick
- 1/4 pc lemon zest
- 5 g extra virgin olive oil
- salt and pepper to judge

METHOD

TOMATO SAUCE

1. Cook the chopped garlic until lightly brown with olive oil in a frying pan over medium heat. Add the onions, continue sweating until they have softened. Add the fresh tomatoes and keep cooking until the mixture has almost dried out.
2. Finally add the tomato paste, tinned tomatoes and chopped basil, cook for another 10 minutes. Season to taste by adding salt and pepper. Set it aside.

TO ASSEMBLE AND COOK

3. Spread some tomato sauce at the bottom of a container. Top up with the mixed vegetables alternatively. Season with salt and pepper.
4. Bake it in the preheated oven at 180°C until all the vegetables are tender, about 15-20 minutes.

TO SERVE

5. Grate the lemon zest and drizzle with olive oil on top, serve it right away.

COOKING TIPS

+ Don't bake the vegetables too long; otherwise they will lose their bright colours.

甜點 | 4 人份

豆漿米布甸配香茅冰沙

Soya milk rice pudding
with lemongrass granita

材料

豆漿米布甸

意大利米	120 克
無糖豆漿	600 毫升
香茅	4 枝
幼砂糖	40 克
椰奶	50 毫升

芒果豆漿泡泡

新鮮芒果漿	100 克
無糖豆漿	70 毫升
淡忌廉	30 毫升

香茅冰沙

水	240 毫升
砂糖	40 克
香茅	4 枝
薑	20 克
檸檬皮	1 個份量
檸檬汁	30 毫升

其他

新鮮芒果（切粒）	2 個份量
焗香杏仁片	30 克
三色紫羅蘭	4 朵

做法

豆漿米布甸

1. 小鍋內下豆漿、砂糖及切碎香茅，煮至微滾，熄火，加蓋靜置 15 分鐘。

2. 豆漿過篩，拌入意大利米煮滾，將混合物倒進深焗盤內，蓋上錫紙。

3. 焗爐預熱至攝氏 180 度，放入焗盤焗 15 至 20 分鐘，拿走錫紙並下椰奶拌勻，靜置冷卻後倒進玻璃杯內，放入雪櫃冷藏。

芒果豆漿泡泡

4. 所有材料放入大碗內拌勻，倒進攪打器，打入二氧化氮製造泡沫，冷藏泡沫。

香茅冰沙

5. 在小鍋內放入所有材料，煮滾，加蓋，並靜置 20 分鐘；混合物過篩，放在冰格冷凍一晚。

6. 用叉刮下冰沙備用。

完成

7. 豆漿米布甸上面放芒果、芒果泡沫、杏仁片及香茅冰沙，用三色紫羅蘭裝飾。

提示

+ 建議先煮好米布甸，再放入玻璃杯內冷藏，味道會更好。

Soya milk rice pudding with lemongrass granita

INGREDIENTS

RICE PUDDING

- 120 g risotto rice
- 600 ml soya milk without sugar
- 4 sticks lemongrass
- 40 g castor sugar
- 50 ml coconut milk

SOYA AND MANGO FOAM

- 100 g fresh mango puree
- 70 ml soya milk without sugar
- 30 ml whipping cream

LEMONGRASS GRANITA

- 240 ml water
- 40 g sugar
- 4 sticks lemongrass
- 20 g fresh ginger
- 1 pc lemon zest
- 30 ml lemon juice

OTHERS

- 2 pcs fresh mango, diced
- 30 g almond slices, toasted
- 4 florets pansy or viola

COOKING TIPS

+ To enhance the flavour, it would be better to cook the rice in advance and and transfer into the glasses to chill before service.

METHOD

RICE PUDDING

1. Combine the soya milk, sugar and chopped lemongrass in a sauce pan and bring them to a simmer. Turn the heat off and cover it with a lid, and let it infuse for 15 minutes.
2. Strain the soya milk and add the risotto rice, bring them to a boil again. And then transfer the mixture into a deep baking tray covered with a piece of foil.
3. Bake it in the pre-heated oven (180°C) for about 15 to 20 minutes or until you are fine with the texture. Remove the foil and finally add the coconut milk, stir to mix. Let it cool and transfer into glasses, chill in the fridge.

SOYA AND MANGO FOAM

4. Combine everything in a mixing bowl and transfer it into a whipper and charge it with N2O gas. Keep it in the fridge for later use.

LEMONGRASS GRANITA

5. Combine everything together in a saucepan and bring it to a boil, cover with a lid and let it infuse for about 20 minutes. Strain it into a container and freeze overnight until hard.
6. Use a fork to scrape some granita, keep them for later use.

TO SERVE

7. Top up the rice pudding with the mango dices, mango foam, toasted almond slices and lemongrass granita. Finally garnish with viola.

豆漿米布甸配香茅冰沙

61

SET/01

summer
menu

帶 子 蔬 菜 沙 律

Scallop ceviche
and vegetable tower

構 思　HOW IS IT DESIGNED?

頭盤 Ceviche 的意思，是將生海鮮浸在橙或檸檬汁內，這樣不但可以保留海鮮的鮮味，還有清新、微酸的味道。我將生帶子、沙律菜用橙汁和檸檬汁醃漬約 5 分鐘，將汁液濾去，並疊成塔型，造型和味道都非常清新吸引。沙律塔可以預早做好再冷藏，可節省時間。主菜是配有薯片的三文魚扒，味道不是很濃，較輕怡，跟牛油粟米很配合。甜品是傳統的法式忌廉蛋白，讓這頓以法國風味作結。

"Ceviche" means raw seafood is cured in citrus juices such as lemon or orange juices. It not only retains the original flavour from the raw ingredients, but also gives the freshness and acidity to the dish. In this case, all the scallops and vegetables are cured in orange and lemon juice for some time, strained and built up to a tower. It is refreshing, vibrant and attractive, the good thing about it is they can be built up in advance and served in an easier way. After that, it is an even fancier dish which is salmon with scales made from potato discs. It is light and matches quite well with the sweet corns which are currently at their peak season. Last but not the least, to finish the menu with the poached egg white served with sauce crème anglaise.

Main Dish
主菜

香 煎 薯 片 三 文 魚
配 牛 油 粟 米

Pan-fried salmon in potato scales
with sweet corns

Dessert
甜點

法 式
忌 廉 蛋 白

Floating island

預 先 準 備 TIPS FOR SMART PREPARATION

+ 將所有蔬菜切片，以供醃漬。

+ 可以先將薯仔切成魚鱗狀，冷藏，用時才取出。

+ 可以預先做好甜品的醬汁，冷藏，用時才取出。

+ Slice all the vegetables and get ready for marinating.

+ Finish preparing the potato scales and keep them in the fridge.

+ Make the sauce for dessert in advance and keep it in the fridge.

頭盤 | 4 人份

帶子蔬菜沙律
Scallop ceviche and vegetable tower

材料

醃 料

橙皮和橙汁	1 個份量
青檸皮和青檸汁	1 個份量
檸檬皮和檸檬汁	1 個份量
初榨橄欖油	50 毫升
鹽、黑胡椒	適量

蔬 菜

溫室青瓜	半條
甘筍	半條
紫色蘿蔔	2 條
新鮮芒果	2 個

其 他

刺身級別帶子（2L）	6 隻
沙律生菜	40 克
三色紫羅蘭	4 朵
碗豆苗	4 條

做法

準 備 醃 漬 料

1. 所有醃漬材料放在大碗內拌勻，靜置 5 分鐘。

醃 漬 蔬 菜 及 帶 子

2. 所有蔬菜切片，並用圓形切模切成與帶子差不多的形狀。

3. 將所有蔬菜和醃料拌勻，醃 5 分鐘，瀝乾蔬菜，醃料留用。

4. 醃料和帶子拌勻，醃 2 分鐘。

蔬 菜 塔

5. 將芒果、帶子、蔬菜片順序放入圓形模內。

6. 冷藏 20 分鐘。

7. 將整個模上碟，移除圓形模。

8. 配以醃料汁和其他配料進食。

提示

不要將蔬菜和帶子醃太久，否則味道會太濃，而且口感會變韌。

STARTER | FOR 4 PORTIONS

Scallop ceviche and vegetable tower

INGREDIENTS

MARINADE

- 1 pc orange juice and zest
- 1 pc lime juice and zest
- 1 pc lemon juice and zest
- 50 ml extra virgin olive oil
- salt and pepper to taste

VEGETABLES

- 1/2 pc green house cucumber
- 1/2 pc carrot
- 2 pcs purple radish
- 2 pcs fresh mango

OTHERS

- 6 pcs scallop, sashimi grade (2L)
- 40 g frisée lettuce
- 4 florets pansy or viola
- 4 pcs pea sprout

COOKING TIPS

+ Remember not to over marinate all the vegetables slices and scallop; otherwise their flavour and texture will be too strong and tough.

METHOD

TO PREPARE THE MARINADE

1. Combine all the ingredients in a mixing bowl and let it sit for about 5 minutes.

TO MARINATE THE VEGETABLES AND SCALLOPS

2. Slice all the vegetables into thin slices and cut them into discs with the same size of scallop by using a cutter.

3. Put all the vegetables into the marinade and let it marinate for about 5 minutes. Strain and keep them for later use.

4. Then add the sliced scallops into the same marinade and let it sit for about 2 minutes.

TO ASSEMBLE THE TOWER

5. Arrange the mango, scallops and vegetable discs into a ring alternatively.

6. Let all the ingredients sit in the ring for about 20 minutes in the fridge.

7. Place the whole ring on a plate and remove it to show the tower.

8. Serve it with extra marinade on the side and other garnishes.

主菜 | 4 人份

香煎薯片三文魚
配牛油粟米

Pan-fried salmon in potato scales with sweet corns

 材料

新鮮去皮三文魚扒 400 克

粟米汁

新鮮甜粟米粒	2 條份量
無鹽牛油	40 克
煮粟米芯的水	200 毫升
全脂牛奶	200 毫升
鹽、黑胡椒	適量

粟米泡沫（或免）

粟米汁	200 毫升
淡忌廉	100 毫升

薯片

美國焗薯	2 個
麵粉	50 克
牛油清	100 克

其他

粟米	1 條份量
無鹽牛油	10 克
鹽、黑胡椒	適量

 提示

可以先灑少許麵粉在三文魚扒上才黏薯片，這樣可令
薯片黏得更牢。

粟米汁和泡沫

1. 平底鑊用小火加熱，下牛油，牛油溶化後倒入粟米粒稍炒 1 分鐘，下粟米芯的水和牛奶煮 10 分鐘，用鹽和黑胡椒調味。

2. 煮好的粟米混合物用攪拌機打成幼滑醬汁，過篩。一半的醬汁用作泡沫用。

薯片

3. 薯仔去皮，切成圓柱體，再切成 1 毫米圓形薄片。薯片撒上薄薄的麵粉，將薯仔彷如魚鱗般排放在鋪上微波爐用保鮮紙的碟上。

4. 放在微波爐裏煮約 2 分鐘，取出待涼。

5. 三文魚扒下少許鹽、黑胡椒調味，將三文魚扒放在薯片上，用刀切成兩份，倒上牛油清，放入雪櫃冷藏。

完成

6. 先煎三文魚扒有薯片的一面，煎至金黃，翻轉三文魚扒，並移離爐火，用平底鑊餘溫煎煮三文魚。

7. 用牛油炒熟粟米，用鹽和黑胡椒調味，盛在碟上，伴三文魚扒和粟米泡沫享用。

香煎薯片三文魚配牛油粟米

MAIN DISH | FOR 4 PORTIONS

Pan-fried salmon in potato scales with sweet corns

INGREDIENTS

- 400 g fresh salmon fillet without skin

SWEET CORN SAUCE

- 2 pcs fresh sweet corn, grains removed
- 40 g unsalted butter
- 200 ml cooking liquor from the corncob
- 200 ml full cream milk
- salt and pepper to taste

SWEET CORN ESPUMA (FOAM/ OPTIONAL)

- 200 ml sweet corn sauce from above
- 100 ml whipping cream

POTATO SCALES

- 2 pcs US baking potatoes
- 50 g plain flour
- 100 g clarified butter

OTHERS

- 1 pc sweet corn
- 10 g unsalted butter
- salt and pepper to taste

METHOD

SWEET CORN SAUCE AND ESPUMA

1. Melt butter in a medium size sauce pan over low heat; sweat the sweet corn without colour for about one minute. Then add the cooking liquor and milk, let it simmer for about 10 minutes. Season with salt and pepper.

2. Transfer the mixture into a blender and blend it into a fine and smooth sauce, pass it through a fine sieve. Keep half for making the foam.

POTATO SCALES

③ Peel the potatoes and cut into cylinder, slice it into about 1 mm thick discs. Lightly coat the discs with flour and arrange them resemble the fish scales on a plate lined with microwave-safe cling film.

④ Cook it in the microwave oven for about 2 minutes. Leave to cool.

⑤ Put the salmon fillets on the potato scales and cut along the sides of the salmon, then pour the clarified butter on top, then let it set in the fridge.

TO FINISH

⑥ Pan fry the salmon scale side down until golden brown, flip it over and remove the pan from the heat and let it finish cooking on the pan.

⑦ Sauté the extra sweet corns with butter. Season with salt and butter. Arrange on a plate. Serve the fillets with the sauce and foam.

COOKING TIPS

Dust the salmon fillets with flour before putting onto the potato scales so as to ensure the potato scales stick better on fish after pan fried.

香煎薯片三文魚配牛油粟米

甜點 | 4人份

法式忌廉蛋白
Floating island

材料

雲尼拿蛋奶醬

全脂牛奶	250 毫升
蛋黃	60 克
雲尼拿豆莢	1 條
幼砂糖	20 克

煮馬令用牛奶

牛奶	300 毫升
幼砂糖	10 克

馬令

蛋白	100 克
幼砂糖	40 克
雲尼拿豆莢	1 條

裝飾

焗香杏仁片	20 克
紅莓	8 粒
開心果	10 克
薄荷葉	4 片

做法

雲尼拿蛋奶醬

1. 雲尼拿豆莢剝開一半，刮出雲尼拿籽。在小鍋中用中火煮滾牛奶和雲尼拿籽，熄火並靜置 4-5 分鐘。

2. 在大碗中，將蛋黃和幼砂糖打勻，逐少倒入熱牛奶混合物，一邊倒入一邊攪拌，避免讓蛋黃煮熟。

3. 將混合物倒回小鍋，用中小火加熱並攪拌 4-5 分鐘，直至醬汁濃稠，可以留在匙背，或醬汁溫度是攝氏 75 度。

4. 將醬汁過篩，待涼，放在雪櫃冷藏。

Floating island

INGREDIENTS

**FOR THE SAUCE
(CRÈME ANGLAISE)**

- 250 ml full cream milk
- 60 g egg yolks
- 1 pc vanilla pod
- 20 g castor sugar

POACHING LIQUOR

- 300 ml milk
- 10 g castor sugar

FOR THE MERINGUE

- 100 g egg whites
- 40 g castor sugar
- 1 pc vanilla pod

GARNISHES

- 20 g toasted almond slices
- 8 pcs raspberries
- 10 g pistachio
- 4 pcs mint leaves

馬令

⑤ 在小鍋中，用小火加熱牛奶和幼砂糖，攪拌直至糖溶解。

⑥ 馬令：在大碗中，用打蛋器打發蛋白至半挺身之後分數次加入糖，打至挺身。

⑦ 將兩隻匙羹浸一浸熱水，用一隻匙羹舀起適量已打發的馬令，利用兩隻匙羹將馬令造成橢圓形，放入熱牛奶內，慢慢煮熟馬令（約1分鐘）

⑧ 馬令煮熟後，放在焗盤上瀝乾水分。

完成

⑨ 將馬令放在碟內，倒入適量雲尼拿蛋奶醬，伴上裝飾享用。

提示

＋ 步驟6：建議打發蛋白至濕性發泡，才加入糖繼續打發。

＋ 煮馬令時不要煮滾牛奶，否則令蛋白過分膨脹破壞橢圓形的外觀。

CRÈME ANGLAISE

1. Split the vanilla pod lengthways and scrape out the seeds. Boil the milk and vanilla seeds in a saucepan over a medium heat. Turn off the heat and let it stand to infuse for about 4-5 minutes.

2. Whisk together the egg yolks and sugar in a mixing bowl, pour the hot milk mixture into the egg mix, a little at a time, so that the eggs do not start to cook.

3. Return the mixture to the sauce pan over a medium low heat and stir continuously for 4-5 minutes, or until the mixture has thickened enough to coat the back of a spoon or a thermometer reads around 75℃.

4. Strain the sauce through a fine sieve into a bowl, leave to cool and then refrigerate.

TO COOK THE MERINGUE

5. Combine the milk and sugar in a saucepan over low heat, stirring to dissolve the sugar.

6. For the meringue, using an electric hand whisk, whisk the egg whites in a bowl until stiff peaks with several addition of sugar.

7. Using a serving spoon dipped in hot water, shape big quenelles of the meringue and gently poach in the milk, turning regularly to ensure they are cooked on both sides (about 1 minute).

8. When fully cooked, gently place on baking tray to drain.

TO FINISH

9. Serve the quenelles of cooked meringue on a soup bowl with generous pool of the chilled crème anglaise and other garnishes.

COOKING TIPS

+ Whisk the egg white until soft peak before adding sugar in step (6).

+ Make sure the milk is not boiling while poaching the meringue, or else it will expand too much and ruin the shape.

法式忌廉蛋白

SET/02

summer menu

煙三文魚蟹肉卷配 蜂蜜蒔蘿蛋黃醬

Smoked salmon and crab meat rolls with honey dill cream

構思 HOW IS IT DESIGNED?

再次以海鮮為套餐的主材料，選用有營養和味道清新的蟹肉和紅鯔魚。

頭盤以煙三文魚的鹹香帶起蟹肉的清鮮；車厘茄乾和西班牙辣香腸味道比較香濃，配合肉質較軟和呈瓣狀的紅鯔魚為主菜就最適合不過。最後以鬆脆的 filo pasrty 作盛器伴新鮮的草莓，為這個套餐增添甜美的餘韻。

Again, it is seafood menu which consists of both crab meat and red mullet. They are both healthy and light in flavour. The saltiness from smoked salmon can complement the lightness of crab meat. The flesh of red mullet itself is soft and flaky, this matches really well with a tangy flavour from a mixture of oven dried tomatoes and chorizo. Finally, the crunchiness from the fillo pastry cup and the freshness of the strawberries give a pleased full stop to the whole menu.

Main Dish

主菜

Dessert

甜點

煎 紅 鰡 魚 配 西 班 牙
香 腸 雜 菜

Pan fried red mullet with cherry tomatoes,
chorizo and fennels

玫 瑰 忌 廉
草 莓 脆 餅

Crunchy strawberry cup
with rose cream

預 先 準 備　TIPS FOR SMART PREPARATION

+ 煙三文魚蟹肉卷和醬汁可以預先做好
放入雪櫃冷藏。

+ 鬆脆的 Filo pasrty 盛器可以預早焗
好，放在密封容器內可儲存 3 天。

+ 車厘茄可以用焗爐焗乾，然後浸在初
榨橄欖油內冷藏，可保鮮一星期。

+ It would be much better to finish the
smoked salmon rolls and let them set
in the fridge. The sauce can be made in
advance and kept in the fridge.

+ The crunchy cup can be finished baking
before and kept in an air-tight container
for up to 3 days.

+ The cherry tomatoes can also be oven
dried and covered with extra virgin olive
oil, it can be kept in the fridge for up to a
week.

77

頭盤 | 4 人份

煙三文魚蟹肉卷配蜂蜜蒔蘿蛋黃醬

Smoked salmon and crab meat roll with honey dill cream

材料

蜂蜜蛋黃醬

原味蛋黃醬	100 克
蜂蜜	10 克
法式芥末	10 克
橄欖油	10 克
新鮮刁草（切碎）	5 克
凍滾水	適量
鹽、黑胡椒	適量

煙三文魚蟹肉卷

煙三文魚片	16 片
罐頭蟹肉	200 克
蜂蜜蛋黃醬	30-40 克
檸檬汁和皮	1/4 個
意大利芫茜（切碎）	5 克
初榨橄欖油	5 毫升
鹽、黑胡椒	適量
牛油果（切片）	2 個

裝飾

三文魚子	10 克
沙律菜	20 克
新鮮刁草	適量

做法

蜂蜜蛋黃醬

❶ 將所有材料拌匀。

煙三文魚蟹肉卷

❷ 蟹肉過篩,用廚房紙印乾水分,與蜂蜜蛋黃醬、檸檬汁、檸檬皮、意大利芫茜、橄欖油同放入大碗內拌匀,下鹽和黑胡椒調味,拌匀。

❸ 在枱面上放雙層保鮮紙,將煙三文魚片排成一大長方形。

❹ 將一半蟹肉混合物舀在三文魚片的中間位置成一直線,排上牛油果片,再放上餘下的蟹肉混合物,將煙三文魚片捲起成三文魚卷,分別將頭和尾的保鮮紙扭實。

❺ 放在雪櫃冰格冷凍 15 分鐘。

完成

❻ 拆開保鮮紙,將呈半冰狀態的煙三文魚卷切成 8 厘米長小段。

❼ 三文魚卷上碟,放上三文魚子、沙律菜和新鮮刁草裝飾,伴蜂蜜蛋黃醬享用。

提示

+ 將煙三文魚卷放在冰格冷凍,會更容易切成整齊的小段。

Smoked salmon and crab meat roll with honey dill cream

INGREDIENTS

FOR THE HONEY DILL DRESSING

- 100 g mayonnaise neutral
- 10 g honey
- 10 g Dijon mustard
- 10 g olive oil
- 5 g fresh dill, chopped
- drinking water to adjust
- salt and pepper to taste

SMOKED SALMON AND CRAB MEAT ROLL

- 16 slices smoked salmon, pre-sliced
- 200 g tinned crab meat
- 30-40 g dressing from above
- 1/4 pc lemon juice and zest
- 5 g Italian parsley, chopped
- 5 ml extra virgin olive oil
- salt and pepper to taste
- 2 pcs avocado, sliced

GARNISHES

- 10 g fresh salmon fish roe
- 20 g mixed leaves
- fresh dill

METHOD

TO MAKE THE HONEY DILL DRESSING

1. Mix everything in a mixing bowl, stir to mix.

SMOKED SALMON AND CRAB MEAT ROLL

2. Strain the crab meat and pat dry with kitchen paper. Combine the crab meat, dressing, lemon juice and zest, chopped parsley and olive oil in a mixing bowl. Season with salt and pepper, stir to mix.

3. Place a double layer of cling film on a working table, neatly put on the pre-sliced smoked salmon into rectangular shape.

4. Spoon half of the crab meat mixture in the middle of the smoked salmon in a straight line. Arrange the sliced avocado on top of crab meat mixture. Cover with the remaining crab meat mixture. Roll it into a log shape and make a tie at both ends.

5. Keep it in the freezer for about 15 minutes before cutting.

TO FINISH

6. Cut the salmon roll (semi-frozen) into about 8 cm long and remove the cling film.

7. Arrange them on a long plate, garnish with salmon roe and baby leaves on top, sauce on the side.

COOKING TIPS

+ Freeze the salmon rolls for a while for easier and neater cutting.

煙三文魚蟹肉卷配蜂蜜蒔蘿蛋黃醬

主菜│4人份

煎紅鯔魚
配西班牙香腸雜菜

Pan fried red mullet
with cherry tomatoes, chorizo and fennels

材料

紅鯔魚

紅鯔魚扒（每件 150 克）	600 克
橄欖油	30 毫升
蒜肉（壓碎）	1 粒
牛油	10 克

焗車厘茄

車厘茄	300 克
蒜片	30 克
新鮮羅勒（切碎）	30 克
橄欖油	30 毫升
意大利黑醋	10 毫升
鹽、糖和黑胡椒	適量

雜菜

橄欖油	30 毫升
西班牙辣香腸（切粒）	100 克
茴香籽	5 克
蒜片	20 克
橙皮	1/3 個
茴香頭（切方粒）	200 克
鹽、黑胡椒	適量

裝飾

羅勒葉	4 片
豌豆苗	少許
法國芫茜、刁草	少許
初榨橄欖油	10 毫升

做法

車厘茄乾

❶ 車厘茄切半，在大碗中與其他材料拌勻醃味，排放在錫紙上。

❷ 放在已預熱攝氏 140 度焗爐中焗半小時，直至車厘茄乾身，保溫。

雜菜

❸ 用平底鑊鑲橄欖油炒香西班牙辣香腸、茴香籽和蒜片，加入茴香頭和橙皮，煮至材料軟化，下鹽和黑胡椒調味，保溫。

紅䱽魚

④ 紅䱽魚扒以鹽和黑胡椒調味，用橄欖油、牛油和蒜碎一起煎至魚皮金黃，翻轉紅䱽魚扒並移離爐火，用平底鑊餘溫煎煮。

完成

⑤ 雜菜加熱，並加入車厘茄乾，拌勻。

⑥ 雜菜上碟，紅䱽魚扒放在上面，用羅勒葉、刁草等裝飾，並下初榨橄欖油供食。

提示

＋ 為保持魚皮脆身，煎前要確保抹乾魚塊，尤其是魚皮。

＋ 臨享用前將雜菜加熱，車厘茄乾要最後才放，否則車厘茄乾會變成糊狀。

Pan fried red mullet
with cherry tomatoes, chorizo and fennels

INGREDIENTS

FOR FISH

- 600 g red mullet fillet, around 150 g/pc
- 30 ml olive oil
- 1 clove garlic, crushed
- 10 g butter

OVEN DRIED CHERRY TOMATOES

- 300 g cherry tomatoes
- 30 g garlic, sliced
- 30g fresh basil, chopped
- 30 ml olive oil
- 10 ml balsamic vinegar
- salt, sugar and pepper to adjust

MIXED VEGETABLES

- 30 ml olive oil
- 100 g chorizo, cut into dices
- 5 g fennel seeds
- 20 g garlic slices
- 1/3 pc orange zest
- 200 g fennel bulbs, cut into cubes
- salt and pepper to taste

FOR GARNISHES

- 4 pcs basil leaves
- pea sprouts
- chervil
- dill
- 10 ml extra virgin olive oil

METHOD

OVEN DRIED CHERRY TOMATOES

1. Cut the tomatoes into half and marinate with other ingredients in a mixing bowl. Then transfer them onto a baking tray lined with foil.

2. Slow-bake in the pre-heated oven at 140°C for about half an hour or until they are dried enough. Keep warm.

FOR THE VEGETABLES

3. Sauté the chorizo, fennel seeds and garlic slices with olive oil, add the fennel bulbs and orange peels. Continue cooking until they are tender, season to taste by adding salt and pepper. Keep warm.

FOR THE FISH

4. Season the fish with salt and pepper, pan fry with olive oil, butter and garlic until golden brown on the skin side. Flip it over and remove the pan from the heat, let the hot pan finish cooking.

TO FINISH

5. Reheat the vegetables and add the baked cherry tomatoes. Toss to mix.

6. Arrange the vegetable mix on a plate and put the fish on top, garnish with leaves and drizzle with extra virgin olive oil.

COOKING TIPS

+ To crispy up the skin of fish, make sure the fish is super dry, especially the skin.

+ While cooking the vegetable mix for the fish, make sure the oven dried tomatoes are added at the last minute, otherwise they will become mushy easily.

煎紅鯔魚配西班牙香腸雜菜

85

甜點 | 4 人份

玫瑰忌廉草莓脆餅
Crunchy strawberry cup with rose cream

 材料

脆 餅 杯

菲路餅皮（filo pastry）	3 張
無鹽牛油（已溶）	40 克
杏仁或榛子粉	40 克
幼砂糖	30 克

玫 瑰 忌 廉

淡忌廉	200 毫升
雲尼拿豆莢（剝開，刮籽）	1 條
玫瑰水	適量
糖霜	20 克

裝 飾

草莓	8 粒
草莓醬	60 克
焗香杏仁片	30 克
薄荷葉	數棵

做法

脆 餅 杯

1. 每張菲路餅皮裁成 4 小張；鋪 1 小張餅皮在工作檯上，塗上牛油，灑下幼砂糖和杏仁粉，重複以上步驟兩次，將餅皮放在餅模內。

2. 放入已預熱攝氏 160-170 度的焗爐內焗 10-15 分鐘，直至金黃，待涼。

玫 瑰 忌 廉

3. 在大碗中打起淡忌廉和糖霜，下雲尼拿籽和玫瑰水拌勻，將忌廉放入袋中，配以星形嘴。

完 成

4. 草莓切粒及切片，備用。

5. 將草莓粒放入脆餅杯內，擠入玫瑰忌廉，飾上草莓片和其他裝飾。

6. 將草莓醬倒進碟內，放上脆餅杯，即可享用。

提示

+ 在享用前才將玫瑰忌廉和草莓舀進脆餅杯內，否則脆餅杯會很快變腍。

+ 你亦可將脆餅杯做成圓形。

Crunchy strawberry cup with rose cream

INGREDIENTS

FILO CUPS

- 3 sheets fillo pastry
- 40 g unsalted butter melted
- 40 g almond/hazelnut powder
- 30 g castor sugar

ROSE CREAM

- 200 ml whipping cream
- 1 pc vanilla pod, split, seeds scraped
- rose water to taste
- 20 g icing sugar

GARNISHES

- 8 pcs fresh strawberry
- 60 g strawberry puree
- 30 g almond slices, toasted
- a few mint leaves

COOKING TIPS

+ Make sure the cream and the strawberries are arranged in the shells before service, otherwise they will go soggy easily.

+ It can be moulded into round shape.

METHOD

FILO CUPS

1. Cut each filo pastry into four pieces. Lay one piece of filo pastry on a working panel, brush it with melted butter and sprinkle with sugar and almond powder. Repeat the same process for 2 times. Transfer the pastry onto a mould.

2. Bake them in the pre-heated oven (160-170°C) for about 10-15 minutes or until golden brown. Allow to cool.

ROSE CREAM

3. Mix cream and icing sugar in a mixing bowl, whisk it until stiff. Stir in the vanilla seeds and rose water. Transfer the cream to a piping bag fitted with star nozzle.

TO FINISH

4. Cut the strawberries into dices and slices. Keep them aside.

5. Arrange the strawberry dices into the cups, and pipe in the cream and top up with the sliced strawberries and other garnishes.

6. Pour the strawberry puree into the serving plate and top up with the filled cups, and serve immediately.

玫瑰忌廉草莓脆餅

SET/03

summer menu

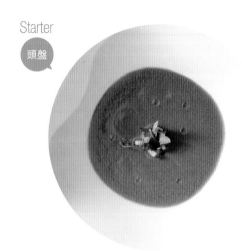

西班牙凍湯

Gazpacho

構思 HOW IS IT DESIGNED?

凍湯,是西班牙傳統的夏日湯品,它的做法非常簡易,可以預早一晚做好並冷藏,味道會更加濃郁。鼠尾草和南瓜是完美的配搭;我以意大利飯為主菜,原因是濃稠、香滑的南瓜和芝士,可平衡入口微辣、清新的凍湯。以經典的甜桃配雲尼拿雪糕作餐後甜品,是圓滿的一餐。

Gazpacho, it is a classic cold soup from Spain served in the summer time. It is easy and simple to make. It can be finished one night in advance and chill it as cold as you can to lift up the flavour. Sage and pumpkin is the perfect match, the risotto is set for the main course because the creaminess from the pumpkin and cheese will bring you to a high level of mouthful satisfaction after the sharp and refreshing flavour from the soup. Finally, peach Melba, which is a very classic and textured dessert, can round up the whole menu successfully.

Main Dish

主菜

鼠 尾 草 芝 士
南 瓜 意 大 利 飯
Pumpkin risotto with sage and cheese

Dessert

甜點

法 式 甜 桃 配
雲 尼 拿 雪 糕
Peach Melba

預 先 準 備 TIPS FOR SMART PREPARATION

+ 可將所有西班牙凍湯的材料拌勻並浸一夜，第二天才用攪拌機攪爛。

+ 在烹調意大利飯的當天，可預先煮意大利米 10 分鐘，讓它涼後，放入雪櫃，待用時取出。

+ 雪糕可預先炮製，放在冰格，享用前才取出。

+ For making Gazpacho, all the ingredients have to be mixed together and let them macerate for one night and blend well the next day.

+ The risotto rice can be pre-cooked for about 10 minutes, chill down right away and keep in the fridge for later use during the day.

+ The ice cream can also be finished and kept in the freezer in advance.

頭盤｜4人份

西班牙凍湯
Gazpacho

材料

湯

熟番茄（切角）	1 公斤
紅西椒（切件）	1 個
黃西椒（切件）	1 個
溫室青瓜（200 克）	1 條
蒜肉（壓碎）	2 粒
羅勒葉（切碎）	20 克
番茄汁	400 毫升
橄欖油	150 毫升
（75 毫升普通＋75 毫升初榨）	
茄膏	80 克
雪莉酒	60 毫升
白酒醋	40 毫升
檸檬汁	適量

調味

鹽、黑胡椒	適量
辣椒仔辣汁	數滴
有汽水	適量

配料

青瓜（切粒）	100 克
新鮮羅勒葉	10 克

 提示

＋ 進食前可以先冷藏湯碗。

 做法

❶ 將所有湯料放入大碗內，徹底拌勻，加蓋，並冷藏過夜。

❷ 翌日用攪拌機將混合物攪至幼滑，過篩，加適量有汽水調校混合物濃度，加適量調味料，須試味。

❸ 伴上青瓜粒和羅勒葉享用。

Gazpacho

INGREDIENTS

SOUP

- 1 kg ripe tomatoes, cut into quarters
- 1 pc red bell pepper, cut into chunks
- 1 pc yellow bell pepper, cut into chunks
- 1 pc (200 g) green house cucumber
- 2 pcs garlic clove, crushed
- 20 g basil leaves, chopped
- 400 ml tomato juice
- 150 ml olive oil (pomace + extra virgin, half and half)
- 80 g tomato paste
- 60 ml sherry wine
- 40 ml white wine vinegar
- lemon juice to taste

SEASONING

- salt and pepper to taste
- a few drops tabasco
- sparkling water to adjust

OTHERS

- 100 g cucumber, diced
- 10 g fresh basil for garnish

METHOD

1. To make the soup, combine everything together in a mixing bowl. Mix them thoroughly, cover and leave it preferably overnight in the fridge.

2. The next day, transfer the mixture into a blender and blend until smooth. Pass it through a fine sieve, adjust the consistency by adding the sparkling water. Season to taste by adding the seasoning.

3. Garnish with cucumber dices and fresh basil.

COOKING TIPS

+ Chill the bowls or cups before serving the soup.

主菜 | 4人份

鼠尾草芝士南瓜意大利飯

Pumpkin risotto with sage and cheese

材料

南瓜蓉

南瓜（去籽切件）	600 克
無鹽牛油	50 克
牛奶	200 毫升
鹽、黑胡椒	適量

意大利飯

意大利米	160 克
白洋葱（切碎）	100 克
無鹽牛油	30 克
白酒	80 毫升
蔬菜上湯或水	1.5 公升
南瓜粒（經牛油稍煎）	100 克
巴馬臣芝士碎	50 克
無鹽牛油（作 monte 之用）	30 克
芝麻菜	40 克
鹽、黑胡椒	適量

裝飾

脆鼠尾草（看 p.196）	數棵
巴馬臣芝士片	20 克
初榨橄欖油	適量

做法

南瓜蓉

❶ 牛油放在小鍋內用中火加熱，加入南瓜並不斷攪拌，煮約 20 分鐘或直至南瓜變軟。

❷ 加入牛奶，並多煮 5 分鐘或至南瓜乾身，下鹽和黑胡椒調味，放進攪拌機內打至幼滑，備用。

意大利飯

❸ 用牛油炒洋葱至透明，加入意大利米並煮約 2 分鐘，灒下白酒拌勻。

❹ 慢慢分多次下上湯或水，不斷攪拌，用中火煮約 16-18 分鐘，直至意大利米熟透。

❺ 加入已經用牛油稍煎的南瓜粒，拌勻，下巴馬臣芝士、牛油及芝麻菜，下鹽和黑胡椒調味。

完成

❻ 飾上芝士、鼠尾草，澆下橄欖油即可享用。

提示

┼ 下牛油和芝士後，不要再加熱煮飯，否則會破壞飯的質感。

Pumpkin risotto with sage and cheese

PUMPKIN PUREE

- 600 g pumpkin,
  deseeded and cut into pieces
- 50 g unsalted butter
- 200 ml milk
- salt and pepper to taste

RISOTTO

- 160 g arborio risotto rice
- 100 g white onion, chopped
- 30 g unsalted butter
- 80 ml white wine
- 1.5 liters vegetable stock or water
- 100 g pumpkin dices, sautéed in butter
- 50 g parmesan cheese, grated
- 30 g unsalted butter (for monte)
- 40 g rocket leaves
- salt and pepper to taste

GARNISHES

- sprigs crispy sage, see p.196
- 20 g parmesan cheese, shaving
- extra virgin olive oil

COOKING TIPS

+ Don't cook the rice again once the butter and cheese are added otherwise it would ruin the texture and taste.

METHOD

PUMPKIN PUREE

1. Heat the butter in a saucepan over medium heat. Add the pumpkins, stir constantly, and cook for about 20 minutes or until they are soft.

2. Add milk and continue cooking for another 5 minutes or until dry. Season to taste with salt and pepper. Transfer the mixture into a blender and blend until smooth, keep it aside.

TO COOK THE RISOTTO

3. Sweat the chopped onions with butter until transparent. Add the rice and continue cooking for a further 2 minutes. Then deglaze with white wine.

4. Start adding the stock or water bit by bit, stirring continuously, and cook the rice for about 16-18 minutes over medium heat or until they are just cooked.

5. Add the sautéed pumpkin dices, stir to mix. Finally add the cheese and Monte with extra butter, rocket leaves. Season to taste with pepper and salt.

TO FINISH

6. Garnish with cheese shaving, crispy sage and olive oil.

鼠尾草芝士南瓜意大利飯

97

甜點 | 4人份

法式甜桃配雲尼拿雪糕
Peach Melba

甜 桃

桃	4 個
橙汁	250 克
檸檬汁	60 克
蜂蜜	50 克
砂糖	40 克
白酒	130 克
黑胡椒粒（壓碎）	6 粒
新鮮百里香	數棵

雲尼拿雪糕

淡忌廉	150 毫升
全脂牛奶	150 毫升
雲呢拿豆莢	1 條
蛋黃	40 克
幼砂糖	50 克

脆 片

低筋麵粉	90 克
無鹽牛油	60 克
幼砂糖	30 克
杏仁片	30 克

裝 飾

新鮮紅莓	100 克
薄荷葉	數片
開心果（磨碎）	5 克

做法

甜 桃

1. 桃去皮，切開一半。

2. 將其餘材料一起煮成糖漿狀，放入桃，並用慢火煮至桃開始變軟，關火，宜將桃浸在糖漿裏一夜。

3. 將桃取出，切片；將糖漿加熱煮至濃稠。

雲尼拿雪糕

4. 牛奶和淡忌廉放在小鍋內加熱，煮至微滾；剝開雲呢拿豆莢，刮出雲尼拿籽，放入牛奶忌廉混合料內，靜置 15 分鐘。

5. 蛋黃和幼砂糖放在大碗內拌勻，慢慢將牛奶忌廉混合料注入蛋黃糊內（邊注入邊攪拌），攪勻後將混合物倒進小鍋，邊攪拌邊加熱至約攝氏 75 度。

6. 將混合物過篩，並用冰水浴冷卻，倒進雪糕機內攪拌至硬。

脆 片

7 將所有材料放入大碗內拌勻，搓成麵糰，放在雪櫃雪至變硬；焗盤鋪上烘焙紙，麵糰削成一片片放在焗盤上，放入已預熱攝氏 180 度的焗爐內焗約 15 分鐘，或至脆片顏色金黃。

完 成

8 將模具放在碟上，用小匙將脆片舀起放入模具內，移開模具，放上桃片及雲尼拿雪糕，澆下煮桃的糖漿。

9 以紅莓、開心果碎和薄荷葉裝飾，立即享用。

提示

+ 宜用慢火煮雪糕混合物（做法 5），並在煮時不斷攪拌，否則會很容易將混合物煮熟至凝固。

💙 DESSERT | FOR 4 PORTIONS

Peach Melba

INGREDIENTS

PEACH COMPOTE

- 4 pcs peach
- 250 g orange juice
- 60 g lemon juice
- 50 g honey
- 40 g sugar
- 130 g white wine
- 6 pcs black peppercorn, crushed
- a few sprigs fresh thyme

VANILLA ICE CREAM

- 150 ml whipping cream
- 150 ml full cream milk
- 1 pc vanilla pod
- 40 g egg yolk
- 50 g castor sugar

CRUMBLE

- 90 g cake flour
- 60 g unsalted butter
- 30 g castor sugar
- 30 g almond slices

GARNISHES

- 100 g fresh raspberries
- a few mint leaves
- 10 g icing sugar, for dusting
- 5 g pistachio, grated

PEACH COMPOTE

1. Peel the peaches and cut them into halves.

2. Boil the rest of the ingredients to syrup and add the peach halves and let it simmer until the peaches are soft but still al dente. Preferably leave the peaches in the syrup overnight.

3. Remove the peaches and cut into slices, reduce the cooking liquor to syrupy consistency.

VANILLA ICE CREAM

4. Pour the milk and cream in a saucepan and bring it almost to the boil. Split open the vanilla pod and scrape out the seeds into the cream mix, let it infuse for about 15 minutes.

5. Mix egg yolk and sugar in a mixing bowl, pour slowly the cream mixture into the egg yolk mixture, stir to mix. Transfer the mixture back to the pan. Stirring continuously, cook until the temperature reached around 75°C.

6. Strain the mixture and cool down over ice water. Then transfer it into an ice cream machine and churn until firm.

CRUMBLES

7. Mix everything together in a mixing bowl, rub to form dough. Chill it until firm, grate it onto a baking tray lined with baking paper. Bake it in the pre-heated oven (180°C) for about 15 minutes or until golden brown.

TO FINISH

8. Mould the crumbles on the plates and top up the peaches and ice cream, spoon the sauce around.

9. Garnish with raspberries, grated pistachio and mint, serve immediately.

COOKING TIPS

+ Cook the ice cream mixture over low heat, stirring continuously by using a spatula, otherwise it would curdle easily.

法式甜桃配雲尼拿雪糕

SET/01

autumn menu

鮮 蝦
忌 廉 湯

Shrimp Bisque

構 思   HOW IS IT DESIGNED ?

這個海鮮套餐,特別為喜歡甲殼類海鮮的粉絲而設。頭盤和主菜都是經典而矜貴的菜式,但不需要花很多時間準備、烹調,非常適合款待摯友回家歡聚。宜在宴客當天購買及烹調海鮮,否則鮮味會很快流失。因為頭盤和主菜的味道頗濃郁,所以我設計一個有朱古力橘子香的甜點,巧用它的甜和酸中和海鮮的濃郁味道。

It is seafood menu, it is good for those who are really indulged in shellfish. Both the starter and main course are classic but luxurious dishes which don't require too much time to prepare. They are must-eat dishes while you are going to invite your closed friends for dinner. For the seafood dishes, ideally they should be prepared and consumed on the day otherwise they will lose their fresh flavour while the time goes on. Although both starter and main course are quite rich, the sweetness and acidity from the chocolate and the citrus can cut off the heaviness at the end.

Main Dish

主菜

法式焗龍蝦
配手切薯條

Lobster thermidor
with hand cut chips

Dessert

甜點

黑朱古力慕絲伴
檸檬忌廉

Dark chocolate mousse
with lemon and lime cream

預先準備 TIPS FOR SMART PREPARATION

+ 甜點可以預早一天做好，放雪櫃冷藏。

+ 焗龍蝦的蘑菇蛋黃醬也可以預早一天
  做好。

+ 手切薯條的薯仔可以預早烚熟，不用加
  蓋並放入雪櫃過夜。

+ 鮮蝦忌廉湯所用的蔬菜可以早一天準
  備，進食當天和鮮蝦再煮即可。

+ The dessert can be finished one day in
  advance and kept in the fridge.

+ The sauce for Lobster Thermidor can be
  made one day before.

+ The hand cut chips can be pre-cooked
  and leave them dry out uncovered
  overnight in the fridge.

+ All the vegetables needed for making the
  bisque can be prepared one day before
  and just finish cooking the day after.

頭盤 | 4 人份

鮮蝦忌廉湯
Shrimp Bisque

蝦 料

新鮮蝦	600 克
蒜肉（壓碎）	2 粒
橄欖油	20 毫升
羅勒	1 棵
無鹽牛油	30 克
白蘭地酒	50 毫升

蔬 菜 料

洋葱（切件）	100 克
甘筍（切件）	100 克
橄欖油	15 毫升
番茄（切件）	2 個
茄膏	20 克
魚湯（看 p.191）	500 毫升
淡忌廉	100 毫升
牛油麵粉糊（看 p.195）	30 克
鹽、黑胡椒	適量

裝 飾

蝦肉（取自上面蝦料）	100 克
豌豆苗	4 棵
淡忌廉	50 毫升

提示

+ 牛油麵粉糊可以大量製造，放在冰格內儲存，隨時可用。

+ 當蝦湯過篩後，可以將水注入篩走的蝦殼內，煮成二湯，作烹調蝦上湯時使用。

+ 進食前加入少許白蘭地酒，能增加蝦湯的香味。

做法

準 備 蝦

① 三分一鮮蝦去殼去腸，下橄欖油醃，放在雪櫃備用（作裝飾用）。

② 餘下的鮮蝦用較剪剪成小件，瀝乾多餘水分。

製 作 蝦 湯

③ 大火燒熱橄欖油，下蒜肉，將蝦煎至微黃，加入牛油和羅勒煮約 1 分鐘，灑下白蘭地酒。

④ 用中火燒熱湯鍋，下橄欖油、洋葱和甘筍煮約 5 分鐘，拌入番茄和茄膏。

⑤ 倒入煎好的蝦和魚湯，慢火煮約 20-25 分鐘。

⑥ 蝦湯過篩，下淡忌廉及牛油麵粉糊，煮至濃稠，下鹽和黑胡椒調味。

完 成

⑦ 用大火快煎備用的蝦肉，盛起放入湯碗內。

⑧ 蝦湯倒進湯碗內，飾上淡忌廉和豌豆苗。

Shrimp Bisque

INGREDIENTS

FOR THE SHRIMPS
- 600 g fresh shrimps
- 2 cloves garlic, crushed
- 20 ml olive oil
- 1 sprig basil herb
- 30 g unsalted butter
- 50 ml brandy

FOR THE VEGETABLES
- 100 g white onion, roughly cut
- 100 g carrot, roughly cut
- 15 ml olive oil
- 2 pcs tomato, roughly cut
- 20 g tomato paste
- 500 ml fish stock, see p.191
- 100 ml whipping cream
- 30 g beurre manie, see p.195
- salt and pepper to taste

GARNISHES
- 100 g fresh shrimps meat from above
- 4 sprigs pea sprout
- 50 ml whipped cream

COOKING TIPS

- + The beurre manie can be prepared in a larger quantity and kept in the freezer until needed.

- + Once the soup has been strained, water can be added to the shrimp shells to make a second stock for the basic stock next time.

- + Add a touch of brandy to the bisque before service will enhance its flavour.

METHOD

TO PREPARE THE SHRIMPS

1. Remove the meat from one-third of the shrimps. De-vein and marinate with olive oil, keep it in the fridge for garnish.

2. Using a pair of scissors, cut the remaining shrimps into small pieces and strain to remove excess water.

TO MAKE THE BISQUE

3. In a frying pan, heat the olive oil over high heat. Add the shrimps together with the crushed garlic, sauté until lightly brown. Then add butter and basil, continue cooking for about 1 minute, finally deglaze the pan with brandy.

4. In a medium size sauce pan over medium heat, add the olive oil, onion and carrot. Cook for about 5 minutes, and then stir in the tomatoes and tomato paste.

5. Transfer the sautéed shrimps into the sauce pan, add the fish stock and let it simmer for about 20-25 minutes.

6. Strain the soup and add the cream, finally thicken the soup with beurre maine.

TO FINISH

7. Quick sauté the shrimps in a frying pan over high heat, transfer them onto the soup bowls.

8. Pour the soup in and garnish with the whipped cream and pea sprout.

主菜 | 4 人份

法式焗龍蝦
配手切薯條

Lobster thermidor
with hand cut chips

 材料

蘑 菇 蛋 黃 醬		手 切 薯 條	
白蘑菇（一開四）	200 克	美國焗薯	2 個
乾葱（切碎）	40 克	鹽、黑胡椒	適量
橄欖油	20 毫升	菜油（炸薯仔用）	1 公升
法式芥末醬	20 克		
白酒	50 毫升	配 料	
魚湯（看 p.191）	200 毫升	沙律生菜	100 克
淡忌廉	200 毫升	芝麻菜	100 克
牛油麵粉糊（看 p.195）	20 克	檸檬汁	10 毫升
蛋黃	2 個	初榨橄欖油	15 毫升
他拉根香草（切碎）	5 克	鹽、黑胡椒	適量
鹽、黑胡椒	適量		

龍 蝦 料	
新鮮波士頓龍蝦	2 隻
橄欖油	50 毫升
白蘭地酒	50 毫升
巴馬臣芝士	60 克

做法

蘑菇蛋黃醬

① 用小火和橄欖油炒乾蔥至軟，倒入蘑菇。

② 煮至蘑菇水分揮發，灒下白酒，注入魚湯。

③ 煮至汁液餘下三分之一，下淡忌廉和芥末醬，煮約 10 分鐘，下牛油麵粉糊煮至濃稠，用適量鹽和黑胡椒調味。

準備龍蝦

④ 切去龍蝦螯，將龍蝦身在滾水裏炻 30 秒，龍蝦螯炻 6 分鐘。

⑤ 龍蝦身切開一半，取去龍蝦腸和眼袋，取龍蝦肉切件；龍蝦螯取肉，抹乾。

手切薯條

⑥ 薯仔洗淨去皮，切成 2x2x8 厘米薯條，在鹽水內滾約 15 分鐘，直至開始變軟。

⑦ 薯條瀝乾水分，抹乾。

完成

⑧ 用大火和橄欖油將龍蝦肉煎至半熟，灒白蘭地酒，下蘑菇醬。

⑨ 移離爐火，下蛋黃、他拉根香草拌匀，盛起舀入龍蝦殼內，灑下巴馬臣芝士，放入攝氏 200 度的焗爐內焗至金黃。

⑩ 同一時間，將菜油加熱至攝氏 180 度，炸薯條至金黃香脆，撈起，下鹽和黑胡椒調味。

⑪ 沙律生菜和芝麻菜放入大碗內，澆下檸檬汁和橄欖油，下鹽和黑胡椒調味。

⑫ 焗好的龍蝦配以薯條和沙律菜享用。

提示

+ 醬汁下蛋黃後不要用火再煮，否則會煮熟蛋黃，結成塊狀。

法式焗龍蝦配手切薯條

Lobster thermidor with hand cut chips

INGREDIENTS

TO PREPARE THE SAUCE
- 200 g white button mushrooms, cut into quarters
- 40 g shallot, chopped
- 20 ml cooking olive oil
- 20 g Dijon mustard
- 50 ml cooking white wine
- 200 ml fish stock, see p.191
- 200 ml whipping cream
- 20 g beurre manie, see p.195
- 2 pcs egg yolks
- 5 g tarragon, chopped
- salt and pepper to taste

TO PREPARE LOBSTER
- 2 pcs Boston live lobster
- 50 ml olive oil
- 50 ml brandy
- 60 g parmesan cheese

HAND CUT CHIPS
- 2 pcs US baking potato
- salt and pepper to taste
- 1 liter vegetable oil for deep frying

OTHERS
- 100 g frisée lettuce
- 100 g rocket leaves
- 10 ml lemon juice
- 15 ml extra virgin olive oil
- salt and pepper to taste

COOKING TIPS
+ Don't cook the sauce again once added the egg yolk otherwise it will be easily curdled.

METHOD

TO PREPARE THE SAUCE
1. Sweat the chopped shallot with olive oil in a sauce pan over low heat until soft, add the mushrooms.
2. Continue cooking until the moisture has evaporated. Deglaze the pan with white wine, add the fish stock.
3. Let it reduce by 2/3 and add the cream and mustard, continue cooking for about 10 minutes, and then thicken it with beurre manie.

TO PREPARE THE LOBSTERS
4. Remove the claws from the lobsters, cook the lobster body and claws in boiling for about 30 seconds and 6 minutes respectively.
5. Cut the body into halves, remove the intestine and eye sac. Remove the meat and cut into pieces; remove the meat from the claws, pat them dry.

HAND CUT CHIPS
6. Wash and peel the potatoes, cut them into about 2x2x8 cm sticks. Cook them in salted simmering water for about 15 minutes (soft but not broken).
7. Strain and pat dry with kitchen paper.

TO FINISH
8. Sauté the lobster meat until half cooked with olive oil in a frying pan over high heat, deglaze with brandy and add the sauce.
9. Remove the mixture from the heat and stir in the egg yolks and chopped tarragon. Transfer the mixture back to the shell and sprinkle with parmesan cheese and grill at 200°C until brown.
10. Meanwhile, deep fry the potato chips in the oil at 180°C until golden brown and crispy, season with salt and pepper.
11. Mix the leaves in a mixing bowl, add the lemon juice, olive oil, season to taste by adding salt and pepper.
12. Serve the grilled lobsters with the crispy chips and salad leaves.

甜點 | 4 人份

黑朱古力慕絲伴檸檬忌廉

Dark chocolate mousse with lemon and lime cream

材料

黑朱古力慕絲
80% 黑朱古力	75 克
蛋黃	60 克約 3 個
幼砂糖	30 克
全脂牛奶	100 毫升
葡萄糖漿	85 克
打起淡忌廉	170 毫升

檸檬青檸忌廉
檸檬汁	50 毫升
雲尼拿豆莢	1/2 條
水	40 毫升
幼砂糖	100 克
蛋白	60 克
打起淡忌廉	125 毫升
青檸皮、檸檬皮茸	各 1 個份量

裝飾
焗香榛子	30 克
朱古力薄片	20 克
乾玫瑰花瓣	10 克
磨碎開心果	5 克

做法

黑朱古力慕絲
1. 黑朱古力放在碗內,浸在熱水浴內至溶,期間要不斷攪拌。
2. 蛋黃和幼砂糖放在大碗內拌勻,轉至小鍋中,下牛奶和葡萄糖漿,邊攪拌邊用小火煮勻。
3. 蛋黃汁會慢慢變濃稠,煮至會在匙羹背上留痕,溫度大約攝氏 80 度,移離爐火並靜置冷卻一會兒。
4. 將 1/3 的蛋黃漿倒入朱古力內,用膠抹刀打圈拌勻,重複以上步驟兩次,之後坐水冷卻。
5. 捲入已打起的淡忌廉,舀入玻璃杯內,放在冰格約半小時。

檸檬青檸忌廉
6. 雲尼拿豆莢剝開一半,刮出雲尼拿籽。將檸檬汁、雲尼拿籽、水和幼砂糖倒入小鍋內,煮成糖漿,溫度大概攝氏 120 度。
7. 用打蛋器將蛋白打至起泡,慢慢倒入糖漿,繼續打混合物直至冷卻至室溫。
8. 下檸檬皮及青檸皮茸,輕力拌勻,再慢慢捲入已打起的淡忌廉。

完成
9. 切碎幾粒焗香榛子,灑在慕絲上,擠上檸檬青檸忌廉,放回冰格雪最少半小時。
10. 這甜點宜在半結冰的狀況下享用,以朱古力薄片、乾玫瑰花瓣和磨碎開心果裝飾。

提示
+ 建議用小火煮蛋漿,並且要不斷攪拌,否則蛋漿會凝固。

Dark chocolate mousse with lemon and lime cream

INGREDIENTS

MOUSSE
- 75 g dark chocolate (80%)
- 60 g (about 3 pcs) egg yolks
- 30 g castor sugar
- 100 ml full cream milk
- 85 g glucose syrup
- 170 ml whipping cream, whipped

LEMON-LIME CREAM
- 50 ml lemon juice
- 1/2 pc vanilla pod
- 40 ml water
- 100 g castor sugar
- 60 g egg white
- 125 ml whipping cream, whipped
- 1 pc for each one lime and lemon zest

GARNISHES
- 30 g hazelnuts, toasted
- 20 g chocolate curls
- 10 g dried rose petals
- 5 g pistachio, grated

COOKING TIPS
+ Use low heat and keep stirring by using a spatula while cooking the custard, otherwise it will curdle easily.

METHOD

DARK CHOCOLATE MOUSSE

1. Melt the chocolate in a mixing bowl over hot water, stirring from time to time.
2. To prepare a pouring custard. In a mixing bowl, beat egg yolks with the sugar, pour this mixture into a sauce pan, add the milk and glucose syrup. Stirring continuously, simmer over low heat.
3. The liquid should thicken slightly and coat the back of a spoon, and the temperature should be around 80°C. Remove it from the heat and cool down slightly.
4. Pour one third of the warm custard into the melted chocolate. Using a spatula, drawing circles to mix. Incorporate the second and third of the liquid, repeat the same procedure. Cool down the mix.
5. Finally fold in the whipped cream. Then transfer the mixture into glass cups and keep them in the freezer for about half an hour.

LEMON-LIME CREAM

6. Split the half vanilla pod lengthways and scrape out the seeds. Cook the lemon juice, vanilla seeds, water and sugar in a sauce pan until the mixture reaches the consistency of syrup. The temperature should read around 120°C.
7. While it is cooking, whisk the egg white with an electric beater until lightly foamy. Pour gradually the hot syrup over the whisked egg white, continuing to beat until the mixture cools down to room temperature.
8. Grate the lemon and lime zest. Add them to the Italian meringue and carefully fold this into the lightly whipped cream.

TO FINISH

9. Chop a few roasted hazelnuts and scatter them on top of the mousse, then pour the lemon-lime cream to cover. Return the cups to the freezer for at least half an hour.
10. It is best served semi-frozen, finally garnish with chocolate curls, rose petals and grated pistachio.

黑朱古力慕絲伴檸檬忌廉

SET/02

autumn menu

菠菜芝士雲吞伴燻肉醬汁

Spinach and ricotta cheese ravioli
with smoked pancetta sauce

構思 HOW IS IT DESIGNED?

這個套餐是「食肉獸」的至愛。
菠菜芝士雲吞伴燻肉醬汁肯定會為你帶來
歡呼，樸質的菠菜與意大利煙燻肉是絕
配。主菜是鮮嫩的慢煮羊柳，每一啖都為
你的客人帶來驚喜。我以紅莓忌廉泡芙作
甜點，酸甜的紅莓能清爽味蕾，以平衡頭
盤、主菜的濃味。

If you don't like seafood, here is another option for
the meat lovers. This simple home-made ravioli kicks
off the menu by giving 'wow' to your friends. The
earthiness of spinach perfectly matches with the
smokiness from the pancetta. Next the tenderness
of the lamb loin will be progressively given
compliments. To finish, the freshness and sourness
from the raspberries can cut off the heaviness
above.

慢煮羊腰肉
配薯蓉砵酒醬汁

Oven slow-cooked lamb loin
with crushed potato and Port wine sauce

紅莓忌廉
泡芙

Paris Brest with raspberry

預先準備 TIPS FOR SMART PREPARATION

+ 雲吞皮麵糰可以預早一天做好，放在雪櫃冷藏，用時才壓成薄皮。

+ 雲吞餡亦可以預早一天做好。

+ 泡芙可以預早一天焗好，放在密封容器裏。伴食的醬汁亦可以預早一天煮好，放在雪櫃冷藏。

+ The pasta dough can be made one day before and kept in the fridge for finishing the next day.

+ The ravioli filling can be finished one day before and used for wrapping the day after.

+ The puffs can be finished one day before and kept at an air-tight container. The sauce can also be finished one day before and kept in the fridge.

頭盤 | 4 人份

菠菜芝士雲吞
伴燻肉醬汁

Spinach and ricotta cheese ravioli
with smoked pancetta sauce

雲吞麵糰

意大利 00 麵粉	150 克
全蛋	25 克
蛋黃	40 克
橄欖油	25 毫升
番紅花水（看 p.195）	25 毫升
鹽	3 克

雲吞餡料

菠菜	120 克
橄欖油	適量
乾葱（切碎）	1 個約 30 克
意大利軟芝士（ricotta）（瀝乾）	120 克
巴馬臣芝士碎（parmesan）	20 克
水牛芝士碎（mozzarella）	20 克
鹽、黑胡椒	適量

醬汁

意大利煙燻肉（切粒）	100 克
乾葱（切碎）	1 個約 30 克
白酒	100 毫升
無鹽牛油	30 克
檸檬汁	10 毫升
番茜（切碎）	5 克

其他

焗脆法國芫茜（看 p.196）	4 棵
全蛋漿（黏雲吞皮用）	30 克

提示

+ 宜只用菠菜葉。
+ 菠菜煮熟後要盡量擠乾水分。

做法

雲吞麵糰

1. 全蛋、蛋黃、橄欖油、鹽和番紅花水倒入大碗內，拌勻。

2. 枱面灑下一層薄麵粉，全部麵粉篩在枱上，在麵粉中央開一個小窩，逐少拌入蛋漿混合物，慢慢將麵粉和蛋漿拌勻。

3. 用手將粗糙粉糰搓成球狀；枱面再灑下一層薄麵粉，將麵糰搓至軟滑及有彈性。

4. 用保鮮紙蓋着麵糰，靜置在室溫半小時，就可壓成薄片及切成食譜所需的形狀。

雲吞餡料

5. 用橄欖油和乾葱炒熟菠菜，盛起瀝乾，切碎，與其他材料拌勻成餡料。將餡料分成 4 份，搓成圓球，在雪櫃冷藏。

醬汁

6. 用中火炒煙燻肉至油分溢出、肉呈微黃（炒約 2 分鐘）。

7. 煙燻肉轉至碟上，用原鑊再炒乾葱約 30 秒。

8. 注入白酒，煮至白酒差不多完全蒸發，移離爐火，倒入煙燻肉及餘下材料，拌勻，下清水或焗意粉水煮至融合。

完 成

⑨ 揉壓麵糰，並用擀麵機壓成 1 毫米厚麵
皮，切成一塊塊 10x10 厘米能包裹餡料的
雲吞皮。

⑩ 將一片雲吞皮放上餡料，在邊緣掃上蛋
液，放上另一片雲吞皮，壓實邊緣成「UFO」
狀；其餘雲吞皮都如法炮製。

⑪ 煲滾淡鹽水，下雲吞煮約 3 分鐘。

⑫ 盛起雲吞，瀝乾水分，放在碟上，舀入暖
醬汁，以脆法國芫茜裝飾。

STARTER | FOR 4 PORTIONS

Spinach and ricotta cheese ravioli with smoked pancetta sauce

INGREDIENTS

RAVIOLI DOUGH
- 150 g Italian 00 flour
- 25 g whole egg
- 40 g egg yolk
- 25 ml olive oil
- 25 ml saffron water, see p.195
- 3 g salt

FILLING
- 120 g spinach
- olive oil to judge
- 1 pc (30 g) shallot, chopped
- 120 g ricotta cheese, drained
- 20 g parmesan cheese, grated
- 20 g mozzarella cheese, grated
- salt and pepper to taste

SAUCE
- 100 g smoked pancetta, diced
- 1 pc (30 g) shallot, chopped
- 100 ml white wine
- 30 g unsalted butter
- 10 ml lemon juice
- 5 g parsley, chopped

OTHERS
- 4 sprigs crisp chervil for garnish
- 30 g whole egg, for sticking pasta sheet

COOKING TIPS

+ Pick the leaves only when prepare the spinach.
+ Squeeze out as much water as possible from the spinach once cooked.

RAVIOLI DOUGH

1. Combine whole egg, egg yolk, olive oil, salt and saffron water together in a mixing bowl, stir well.
2. Sift the flour on a large floured surface. Make a well in the center, add the egg mixture bit by bit, and gently start working the flour into the liquid.
3. Use your hands to form the rough dough into a ball. Transfer the dough to a lightly floured surface and knead it until smooth and elastic.
4. Cover it with cling film and let rest for half an hour at room temperature. Proceed with rolling and cutting the pasta according to the recipe.

FILLINGS

5. Cook the spinach with shallot and olive oil, drained well and chopped. Mix it with the remaining ingredients. Divide the filling into about 4 portions and roll them into a ball. Keep them in the fridge.

SAUCE

6. Sauté the pancetta in a frying pan over medium heat, stirring occasionally, until fat is rendered and pancetta is lightly browned, about 2 minutes.

7. Transfer the pancetta to a plate, add the chopped shallot in the frying pan, keep stirring, and cook for about 30 seconds.
8. Add the white wine and cook until it is almost completely evaporated. Remove from heat and return pancetta to the pan. Add the remaining ingredients and pasta or plain water to emulsify.

TO FINISH

9. Knead the pasta dough and roll it into about 1 mm thick using a pasta machine, cut it into individual squares (10x10 cm) big enough to cover the filling.
10. Place the filling in the center of pasta sheet. Brush the edges of the pasta sheet with a bit of egg wash. Place another pasta sheet over the top of the filled pasta sheet and press down gently around the filling. The shape likes a 'UFO'. Repeat the process with the remaining pasta sheets and fillings.
11. Cook the ravioli in salted boiling water for about 3 minutes.
12. Drain the ravioli and put them on a plate, spoon the warm sauce around and garnish with the crispy chervil.

菠菜芝士雲吞伴燻肉醬汁

119

主菜 | 4 人份

慢煮羊腰肉
配薯蓉砵酒醬汁

Oven slow-cooked lamb loin with
crushed potato and port wine sauce

 材料

羊腰肉	2 件約 400 克
橄欖油	10 毫升
蒜肉（壓碎）	20 克
新鮮百里香	數棵
鹽、黑胡椒	適量

薯蓉

連皮新薯	4-5 個
新鮮百里香葉（切碎）	4 棵份量
初榨橄欖油	30 毫升
鹽、黑胡椒	適量

砵酒醬汁

乾葱（切碎）	20 克
橄欖油	20 毫升
麵粉	5 克
砵酒	200 毫升
雞湯（看 p.190）	100 毫升
鹽、黑胡椒	適量

開心果芝士脆皮

切碎開心果	40 克
巴臣馬芝士碎	15 克
鹽、黑胡椒	適量

提示

+ 鋪上開心果芝士脆
  皮材料前，先在羊
  腰肉塗上一層芥末
  醬，味道會更香。

羊腰肉

1. 切去羊腰肉周邊多餘油脂，用廚房紙抹乾水分，下鹽和黑胡椒調味，用棉繩紮實羊腰肉。燒熱平底鑊，下橄欖油、蒜茸和百里香，放入羊腰肉煎至金黃。

2. 焗架放在焗盤上，將羊腰肉轉放焗架，放入已預熱攝氏110度焗爐內焗12-15分鐘。剪去棉繩。

薯蓉

3. 洗淨新薯，用淡鹽水焓至軟熟，撕去薯皮，壓碎成薯蓉，下其他材料拌勻，保溫。

砵酒醬汁

4. 用橄欖油將乾葱炒至軟，加入麵粉拌勻，灒下雞湯，煮至約餘下三分一。

5. 注入砵酒，並繼續煮至餘下一半，或汁液濃稠至可在匙背留痕，下鹽和黑胡椒調味。

開心果芝士脆皮

6. 所有材料拌勻，備用。

完成

7. 將開心果芝士脆皮材料鋪在羊腰肉面，以攝氏180度焗1分鐘，直至芝士溶化黏實羊柳。

8. 從焗爐取出羊腰肉，靜置5分鐘，切件，配砵酒醬汁和薯蓉享用。

Oven slow-cooked lamb loin with crushed potato and port wine sauce

INGREDIENTS

- 2 pcs (400 g) lamb loin
- 10 ml olive oil
- 20 g garlic, crushed
- sprigs fresh thyme
- salt and pepper to judge

CRUSHED POTATO
- 4-5 pcs new potato with skin on
- 4 sprigs fresh thyme leaves, chopped
- 30 ml extra virgin olive oil
- salt and pepper to taste

PORT WINE SAUCE
- 20 g shallot, chopped
- 20 ml olive oil
- 5 g plain flour
- 100 ml chicken stock, see p.190
- 200 ml port wine
- salt and pepper to taste

PISTACHIO AND CHEESE CRUST
- 40 g pistachio nuts, chopped
- 15 g parmesan cheese, grated
- salt and pepper to taste

COOKING TIPS
+ Brush a little of mustard onto the lamb loin before put the crust ingredients on. It can enhance the flavour.

METHOD

TO PREPARE THE LAMB
1. Trim the lamb loins and pat them dry with kitchen paper, season with salt and pepper. Tie the lamb loins with butcher string. Pan fry until golden brown on a hot pan with olive oil, garlic and thyme.

2. Transfer them onto a wire rack sitting on a tray, cook it in for about 12-15 minutes (for medium rare) in the pre-heated oven of (110°C). Remove and discard the string.

CRUSHED POTATO
3. Cook the cleaned potatoes in the simmering salted water until tender. Skin the potatoes. Crush the potatoes and add the remaining ingredients, mix well. Keep warm aside.

PORT WINE SAUCE
4. Sweat the chopped shallot with olive oil in a sauce pan until soft. Add the flour, stir to mix. Deglaze the pan with chicken stock, reduce it by 2/3.

5. Add the port wine and continue reducing by half or until it is thick enough to coat the spoon. Finally adjust the seasoning by adding salt and pepper.

PISTACHIO AND CHEESE CRUST
6. Combine everything together in a mixing bowl, set it aside.

TO FINISH
7. Then put the crust ingredients on top of lamb loins and put them under the grill (180°C) for about 1 minute or until the cheese has just melted and stuck on the lamb loin.

8. Let it rest for about 5 minutes, then cut it into slices and serve them with the sauce and crushed potatoes.

慢煮羊腰肉配薯蓉砵酒醬汁

123

甜點 | 4人份

紅莓忌廉泡芙

Paris Brest with raspberry

蛋 黃 忌 廉

牛奶（1）	200 毫升
牛奶（2）	50 毫升
幼砂糖	30 克
蛋	1 個
吉士粉	10 克
粟粉	10 克
雲尼拿豆莢	1/2 條
打起淡忌廉	50 毫升
魚膠片（浸透）	5 克

曲 奇 皮

幼砂糖	50 克
無鹽牛油	50 克
麵粉	50 克

泡 芙 粉 漿

無鹽牛油	60 克
水	75 毫升
牛奶	75 毫升
幼砂糖	4 克
鹽	少許
全蛋	150 克
麵粉	96 克

醬 汁

雲尼拿蛋奶醬（看 p.194）	150 毫升

裝 飾

紅莓	16 粒
糖霜	10 克
磨碎開心果	5 克
薄荷葉	4 棵

做法

蛋 黃 忌 廉

① 刮出雲尼拿籽，煮熱牛奶（1），下雲尼拿籽，靜置 10 分鐘，備用。

② 大碗內拌勻幼砂糖、吉士粉、粟粉，在另一個碗拌勻蛋和牛奶（2），分三次倒入乾粉混合物內拌勻。

③ 步驟 1 和 2 的混合物拌勻，在小鍋內過篩，以中火邊煮邊攪拌，煮至濃稠，倒入已浸水的魚膠片，攪拌直至魚膠片溶解。

④ 將混合物轉至盤內，蓋上保鮮紙，放入雪櫃冷卻。

⑤ 取出混合物打至軟滑，加入已打起的淡忌廉，唧進唧袋內，放在雪櫃冷藏。

曲 奇 皮

⑥ 大碗中拌勻所有材料，用手揉勻，將麵糰壓在兩張烘焙紙之間，壓成 2 毫米厚。

⑦ 放在冰格雪至硬，切成泡芙大小一樣的圓片。

紅
莓
忌
廉
泡
芙

125

泡芙粉漿

⑧ 在小鍋內拌勻牛奶、無鹽牛油、水、幼砂糖和鹽，煮滾，拌入過篩麵粉，煮約 1 分鐘。

⑨ 移離爐火，逐個蛋加入（每次都要用打蛋器徹底拌勻，才再加入另一個蛋。）

⑩ 粉漿舀進唧袋，在不黏底墊上唧出圓形圖案，將曲奇皮放在圓形粉漿上，放在已預熱 180-200 度焗爐內焗約 15-20 分鐘，直至金黃香脆。

組合

⑪ 泡芙冷卻後，在底部刺孔，唧入蛋黃忌廉。

⑫ 以紅莓、糖霜、開心果和雲尼拿蛋奶醬裝飾及伴食。

提示

+ 臨近奉客前，泡芙可以再放入焗爐焗脆點。

+ 盡量在進食前才唧蛋黃忌廉，否則泡芙會變濕變脸。

Paris Brest with raspberry

INGREDIENTS

CUSTARD CREAM
- 200 ml milk (1)
- 50 ml milk (2)
- 30 g castor sugar
- 1 pc whole egg
- 10 g custard powder
- 10 g corn starch
- 1/2 pc vanilla pod
- 50 ml whipping cream, whipped
- 5 g gelatin sheet, soaked in cold water

TOPPING
- 50 g castor sugar
- 50 g unsalted butter
- 50 g plain flour

CHOUX PASTE (PROFITEROLES)
- 60 g unsalted butter
- 75 ml water
- 75 ml milk
- 4 g castor sugar
- pinch salt
- 150 g whole eggs
- 96 g plain flour

SAUCE
- 150 ml sauce (crème anglaise), see p.194

GARNISH
- 16 pcs fresh raspberry
- 10 g icing sugar for dusting
- 5 g pistachio nuts, grated
- 4 sprigs mint

CUSTARD CREAM

① Scrape out the seeds from vanilla pod and add them into the hot milk (1) and let it infuse for about 10 minutes.

② Mix castor sugar, custard powder and corn starch in a mixing bowl. In another mixing bowl, combine whole egg and milk (2), then pour into the dry mixture in 3 times until it is fully mixed.

③ Combine the mixture in process (1) and (2), then pass it through a sieve into a sauce pan. Cook the mixture over medium heat, stirring continuously, until it thickens. Add the soaked gelatin, stir to dissolve.

④ Remove it from the heat and transfer onto a tray, cover with cling film and chill it down in the fridge.

⑤ Take it out and whisk until smooth, finally fold in the whipped cream. Transfer it into a piping bag and keep in the fridge for later use.

COOKIE TOPPING

⑥ Combine everything together in a mixing bowl, rub to mix. Roll it out into about 2 mm thick in between 2 parchment paper.

⑦ Freeze it until hard; cut it into the round shape just big enough to cover the puff.

PROFITEROLES

⑧ Mix milk, unsalted butter, water, castor sugar and salt in a sauce pan, bring them to a boil. Stir in the sieved flour, cook it for about a minute.

⑨ Remove it from the heat and add the eggs one by one, whisk to mix thoroughly with electric mixer before adding the other one.

⑩ Transfer the mixture into a piping bag and pipe a wheel shape on to a non-stick mat. Arrange the cookie topping on top of each puff and bake them in the pre-heated oven (180-200°C) for about 15-20 minutes or until they are golden brown and crispy.

TO ASSEMBLE

⑪ Once the profiteroles have cooled down, prick holes at the bottom of each puff and pipe the filling in.

⑫ Finally decorate with the raspberries, icing sugar, grated pistachio nuts and sauce crème anglaise.

> **COOKING TIPS**
>
> + The finished puffs can be re-baked to crisp up before service.
> + Try to pipe in the filling right before service otherwise the puff would go soggy.

紅莓忌廉泡芙

127

SET/03

autumn menu

焦糖洋葱羊奶芝士多士 配芝麻菜沙律

Caramelized onion and goat cheese on
toast with rocket and walnut salad

構思 HOW IS IT DESIGNED ?

在芸芸美食中，素食可能不是首選，但這個素食套餐色香味俱全，尤其是主菜的雜菜藜麥配私房番茄醬汁，層層疊起的蔬菜，色彩繽紛，引起食欲；焦糖洋葱和山羊芝士永遠都是好配搭，這不用多説；並以甜美的棗子拖肥蛋糕配冧酒提子雪糕為整個套餐作結尾。

Vegetarian menu is not always the first priority choice for most diners. This one is healthy and looks attractive, especially the ratatouille with colourful mixed vegetables stacking up together served with home-made tomato sauce. Caramelized onion and goat cheese is always a perfect match. Lastly, the toffee pudding rounds off the menu by giving sweetness to diners.

Main Dish
主菜

Dessert
甜點

雜 菜 藜 麥
配 私 房 番 茄 醬 汁

Vegetables tower with quinoa and tomato sauce

棗 子 拖 肥 蛋 糕

Sticky toffee pudding with dates

預 先 準 備 TIPS FOR SMART PREPARATION

+ 雜菜塔可以預早做好，留在模內，臨享
　用前才扣出來。

+ 棗子拖肥蛋糕可以預先焗好，進餐前放
　入焗爐裏翻熱就成。

+ The ratatouille tower can be finished and left in the ring before service.

+ The pudding can finish baking before service and reheat in the oven during service.

頭盤 | 4 人份

焦糖洋葱羊奶芝士多士配芝麻菜沙律

Caramelized onion and goat cheese on toast
with rocket and walnut salad

材料

裸麥麵包（厚片）	4 塊

多士餡

紫洋葱（切片）	500 克約 2 個
橄欖油	20 毫升
新鮮百里香	2 棵
鹽、黑胡椒	適量
山羊芝士（切片）	200 克

沙律

去皮合桃（焗香壓碎）	50 克
初榨橄欖油	60 毫升
意大利黑醋	5-10 毫升
鹽、黑胡椒	適量
芝麻菜葉（洗淨）	100 克
沙律生菜葉（洗淨）	50 克

做法

多士餡

① 用厚底大鑊及中大火加熱橄欖油，炒洋葱和百里香約 30 分鐘，直至洋葱呈焦糖化，下鹽和黑胡椒調味，離火，保溫。

烤多士

② 麵包放入已預熱攝氏 180 度的焗爐內焗至微黃，加上焦糖洋葱，再鋪上山羊芝士，焗至微溶及金黃。

沙律汁

③ 大碗內拌勻合桃碎、意大利黑醋和橄欖油，下鹽和黑胡椒調味。

完成

④ 沙律菜和沙律汁拌勻，配多士享用。

 提示

+ 可以用法包代替裸麥麵包。
+ 不要用大火炒洋葱，否則會很容易燒焦。

STARTER | FOR 4 PORTIONS

Caramelized onion and goat cheese on toast with rocket and walnut salad

INGREDIENTS

▨ 4 pcs rye bread, cut into thick slices

FOR THE TOPPINGS
▨ 500 g (2 pcs) red onion, sliced
▨ 20 ml olive oil
▨ 2 sprigs fresh thyme
▨ salt and pepper to taste
▨ 200 g goat cheese, cut into thin slices

ROCKET AND WALNUT SALAD
▨ 50 g walnuts, skin off, toasted and crushed
▨ 60 ml extra virgin olive oil
▨ 5-10 ml balsamic vinegar
▨ salt and pepper to taste
▨ 100 g rocket leaves, rinsed
▨ 50 g frisée leaves, rinsed

METHOD

FOR THE TOPPING
1 Sauté the sliced onions with olive oil and thyme until they are nicely caramelized in heavy large skillet over medium-high heat, about 30 minutes. Season to taste by adding salt and pepper. Keep warm for service.

TO GRILL THE TOAST
2 Grill the bread under the pre-heated oven (180°C) until lightly brown. Arrange the caramelized red onion on top and top up with the cheese. Grill the cheese until lightly brown and nearly melted.

TO MAKE THE DRESSING
3 Whisk to combine the walnuts, balsamic vinegar and olive oil in a mixing bowl. Season to taste by adding salt and pepper.

TO FINISH
4 Mix the salad leaves with the dressing and serve it on the side with the toast.

COOKING TIPS
+ Baguette can replace rye bread if you couldn't find it.
+ Don't use high heat to cook onions otherwise they will be burnt easily.

主菜 | 4人份

雜菜藜麥配私房番茄醬汁

Vegetable tower with quinoa and tomato sauce

材料

雜菜塔

黃色意大利青瓜（切片）	150 克
綠色意大利青瓜（切片）	150 克
茄子（切片）	150 克
連莖番茄（切片）	2 個
蒜肉（壓碎）	2 粒
新鮮羅勒葉	4 片
初榨橄欖油	30 毫升
鹽、黑胡椒	適量

藜麥料

紅西椒（切粒）	1/2 個
黃西椒（切粒）	1/2 個
初榨橄欖油	20 毫升
蒜肉（壓碎）	1 粒
藜麥	80 克
水	200 毫升
焗香杏仁片	50 克
檸檬皮和汁	1/3 個
鹽、黑胡椒	適量

番茄醬汁

初榨橄欖油	20 毫升
蒜頭（切碎）	2 個
白洋葱（切碎）	50 克
番茄（切粒）	2 個
茄膏	20 克
罐頭番茄（切碎）	100 克
新鮮羅勒葉（切碎）	10 克

裝飾
豌豆苗

做法

雜菜塔

1. 所有切片蔬菜用橄欖油和蒜頭煎至微黃，將蔬菜梅花間竹地在模內疊起（青瓜、茄子、番茄、羅勒葉，重複一次），每一層灑下橄欖油、鹽和黑胡椒。

2. 蔬菜塔放入已預熱 170 度焗爐內焗 10-15 分鐘，直至所有蔬菜變軟，保溫。

藜麥料

3. 用橄欖油和蒜頭將西椒炒至軟，下鹽和黑椒調味，備用。

4. 小鍋內下水和藜麥，煮滾，轉至小火煮約 10 分鐘，移離爐火，加蓋，直至所有水分吸乾。

5. 進食前，拌入其餘材料。

提示

+ 進食前才將杏仁片加入藜麥料內，否則杏仁片會變脸。

番 茄 醬 汁

⑥ 小鍋內下橄欖油，放入蒜頭粒炒至微黃，下洋葱煮約 3 分鐘。

⑦ 下番茄再煮 5 分鐘，拌入茄膏和罐頭番茄。

⑧ 一邊攪拌一邊用小火滾大約 10-15 分鐘，拌入羅勒。

完 成

⑨ 用匙羹將番茄醬汁舀進湯碟內，藜麥放在湯碟中央，將暖的蔬菜塔小心放在藜麥上面，移走圓形模。

⑩ 灑下初榨橄欖油和飾上豌豆苗享用。

Vegetable tower with quinoa and tomato sauce

INGREDIENTS

VEGETABLE TOWER
- 150 g yellow zucchini, sliced
- 150 g green zucchini, sliced
- 150 g eggplant, sliced
- 2 pcs tomato on the vine, sliced
- 2 cloves garlic, crushed
- 4 pcs fresh basil leaves
- 30 ml extra virgin olive oil
- salt and pepper to taste

QUINOA MIX
- 1/2 pc red bell pepper, diced
- 1/2 pc yellow bell pepper, diced
- 20 ml extra virgin olive oil
- 1 clove garlic clove, crushed
- 80 g quinoa
- 200 ml water
- 50 g almond slices, toasted
- 1/3 pc lemon zest and juice
- salt and pepper to taste

TOMATO SAUCE
- 20 ml extra virgin olive oil
- 2 pcs garlic, chopped
- 50 g white onion, chopped
- 2 pcs tomato, diced
- 20 g tomato paste
- 100 g tinned tomatoes, chopped
- 10 g fresh basil leaves, chopped

GARNISHES
- bean sprout

雜 菜 藜 麥 配 私 房 番 茄 醬 汁

METHOD

VEGETABLE TOWER

1. Pan fry the vegetable slices until lightly brown with olive oil and garlic. Stack the veggies slices in alternating patters (e.g. zucchini, eggplant, tomato, basil and repeat) and place them in a mould. Drizzle with olive oil and season while stacking each layer.

2. Bake it in the pre-heated oven (170°C) for about 10-15 minutes or until all the veggies are tender. Keep it warm.

QUINOA MIX

3. Sweat the peppers with olive oil and garlic until tender, season to taste by adding salt and pepper, remove it from the heat and set it aside.

4. In a sauce pan, combine water and quinoa together and bring them to a boil. Turn the heat down and simmer for about 10 minutes. Take off heat and cover until all the liquid has been absorbed.

5. Stir in the remaining ingredients before service.

TOMATO SAUCE

6. In sauce pan, cook the chopped garlic with olive oil until lightly brown, then add the chopped onions and cook for about 3 minutes.

7. Add the chopped tomatoes and cook for another 5 minutes. Stir in the tomato paste and tinned tomatoes.

8. Let it simmer for about 10-15 minutes, stirring constantly. Finally add the chopped basil.

TO FINISH

9. Spoon the tomato sauce on the soup bowls and the quinoa at the center. Transfer carefully the warm vegetable towers onto the top of quinoa, remove the mould.

10. Drizzle with extra olive oil and garnish with the bean sprout.

COOKING TIPS

+ Mix in the toasted almond slices into the quinoa mix just before service otherwise it would turn soggy.

❤ 甜點 | 4 人份

棗子拖肥蛋糕
Sticky toffee pudding with dates

提示

- 冧酒和提子乾要在雪糕打好後才加入，否則雪糕不會變硬。
- 焗蛋糕前，用錫紙圍着蛋糕模，可以防止粉漿滿瀉。

材料

甜棗蛋糕

篩勻 ⎡ 自發粉	90 克
⎣ 梳打粉	1.5 克
無鹽牛油（軟）	40 克
黃砂糖	70 克
糖漿	15 克
雞蛋（打勻）	50 克
牛奶	50 毫升
甜棗（去核切碎）	110 克
紅茶（例如伯爵茶）	90 毫升

拖肥醬

黃砂糖	70 克
無鹽牛油	25 克
淡忌廉	125 毫升
糖漿	7 克

冧酒提子雪糕

蛋黃	40 克
幼砂糖	50 克
淡忌廉	300 毫升
雲尼拿豆莢	1/2 條
一起浸透 ⎡ 提子乾	50 克
⎣ 冧酒	25 克

其他

糖霜	10 克
乾雲尼拿豆莢	4 條

做法

甜棗蛋糕

1. 軟牛油打鬆，下糖再打勻，蛋漿分 2 至 3 次加入，每次均要徹底拌勻；拌入糖漿。

2. 下 1/3 麵粉混合物，再下 1/3 牛奶拌勻，重複步驟兩次；下切碎甜棗和紅茶拌勻。

3. 蛋糕模搽上一層牛油和麵粉，將混合物倒至 3/4 滿，放入已預熱焗爐，以攝氏 160 度焗 25-30 分鐘直至焗熟。

做法

冧酒提子雪糕

4. 提子乾浸冧酒最少 2 小時，最好浸過夜。

5. 淡忌廉倒入厚底鍋內煮至差不多滾，剝開雲尼拿豆莢，刮下雲尼拿籽，將雲尼拿籽放入淡忌廉內靜置約 15 分鐘，備用。

6. 在大碗內打勻蛋黃和糖至淺色，將淡忌廉慢慢倒入蛋黃混合物內，不斷攪拌。

7. 將混合物過篩到乾淨小鍋裏，邊攪拌邊用小火煮約 10-15 分鐘至濃稠，能在匙背留痕，溫度接近攝氏 78 度。

8. 蛋黃忌廉轉放入大碗內，用冰水浴冷卻。

9. 冷卻後用雪糕機打至厚身，下冧酒和提子乾，放在冰格冷凍至硬。

拖肥醬

10. 牛油、糖和一半的淡忌廉放入鍋內，慢慢煮滾，攪拌直至糖完全溶解，拌入糖漿再煮 2 分鐘，移離爐火冷卻 1 分鐘，再拌入餘下淡忌廉。

完成

11. 蛋糕移離蛋糕模，放入碟上，舀入拖肥醬和雪糕。

12. 灑上糖霜和用雲尼拿豆莢裝飾。

COOKING TIPS

+ Don't add the soaked raisins to the ice cream base before it has churned to thick, otherwise it won't set properly.

+ Wrap the tins up with foil while the cakes are baking, this can prevent the cake mix from spilling out.

Sticky toffee pudding with dates

INGREDIENTS

FOR THE PUDDING
- 90 g self raising flour ⎤
- 1.5 g baking soda ⎦ — sieved
- 40 g soft unsalted butter
- 70 g brown sugar
- 15 g treacle
- 50 g whole egg, beaten
- 50 ml milk
- 110 g dried dates, pitted and chopped
- 90 ml red tea (e.g. Earl Grey)

RUM AND RAISIN ICE CREAM
- 40 g egg yolk
- 50 g caster sugar
- 300 ml whipping cream
- 1/2 pc vanilla pod
- 50 g raisins ⎤
- 25 g rum ⎦

soaked

TOFFEE SAUCE
- 70 g brown sugar
- 25 g unsalted butter
- 125 ml whipping cream
- 7 g treacle

OTHERS
- 10 g icing sugar for dusting
- 4 pcs dried vanilla pod husk

METHOD

FOR THE PUDDING

1. Cream the butter until light and fluffy, add the sugar. Emulsify the eggs in 2 to 3 times, beating well between each addition, then add the treacle and combine.
2. Fold in one third of the flour mixture, then one third of the milk. Repeat each step until the flour and milk is used up. Add the chopped dates and the tea mixture. Then stir until combined.
3. Spoon the mixture into the prepared tins (buttered and floured) around 3 quarters full, bake them in the pre-heated oven (160°C) for about 25 to 30 minutes or until they are cooked.

RUM AND RAISIN ICE CREAM

4. Soak the raisin in rum at least 2 hours, preferably overnight or even longer.
5. Pour the cream in a heavy-based saucepan and bring it almost to the boil. Split open the vanilla pod and scrape out the seeds into the cream, let it infuse for about 15 minutes.
6. Combine the egg yolks and sugar in a mixing bowl and whisk together until pale. Pour slowly the cream into the egg yolk mixture, stirring all the time with a whisk.
7. Strain the mixture into the rinsed-out saucepan and cook over a low heat for 10-15 minutes, stirring all the time, until the mixture thickens enough to coat the back of the spoon or has reached a temperature of around 78°C.
8. Remove the custard from the heat and leave to cool in a mixing bowl over an ice water bath.
9. Transfer the mixture into an ice cream machine and churn until thick, add the soaked raisin and freeze until hard to serve.

TOFFEE SAUCE

10. Melt the butter, sugar, and HALF the cream in a pan, slowly bring to the boil, stirring until the sugar has dissolved. Add the treacle and allow the sauce to continue to bubble for a further 2 minutes, then remove from the heat. Leave to cool for a minute, and stir in the remaining cream.

TO FINISH

11. Remove the cakes from the tins and serve them warm with the sauce and ice cream.
12. Finally dust it with icing sugar and garnish with the vanilla pod husks.

棗子拖肥蛋糕

SET /01

winter
menu

煙 三 文 魚
薑 味 炒 蛋

Smoked salmon scrambled eggs
flavoured with ginger

構 思　　HOW IS IT DESIGNED ?

這個套餐的準備功夫超級簡易，但簡易並不代表平凡，反之非常有驚喜。套餐均衡採用海鮮和肉類，能顧及賓客的口味，讓他們為你的貼心而歡呼。

頭盤煙三文魚薑味炒蛋要在進食前才煮，而肉眼扒只需要輕輕炙兩邊，然後在焗爐裏慢煮。在享用甜點前，才將蛋糕浸在冧酒糖漿內，配忌廉享用。

這套餐實在功夫少但成效大。

This menu is super easy to prepare, but they definitely can give 'wow' factor to your closed friends and guests. This menu is well balanced by incorporating seafood and meat, both of them here don't need to take plenty of time for the preparation. The egg dish can cook at the last minute, and the beef is just seared and put into oven to cook slowly for perfection. To finish the menu, just soak the baba cake into the pre-made syrup and served with the whipped cream.

Main Dish
頭盤

Dessert
甜點

慢 煮 美 國 肉 眼 扒
伴 羊 肚 菌 忌 廉 汁

Oven roasted (slow-cooked style)
US beef ribeye with morel sauce

法 式
冧 酒 蛋 糕

Rum baba

準 備 心 得　TIPS FOR SMART PREPARATION

+ 可以預先準備蛋漿及羊肚菌白汁，放
  在雪櫃冷藏。

+ 蛋糕可以預先做好，放在室溫保
  存；冧酒糖漿則可以放在雪櫃冷藏。

+ The egg mixture and the morel
  sauce can be finished before and
  kept in the fridge.

+ Baba cake and syrup can be finished
  in advance and keep it at room
  temperature and fridge respectively.

頭盤 | 4 人份

煙三文魚薑味炒蛋

Smoked salmon scrambled eggs
flavoured with ginger

材料

炒 蛋

新鮮雞蛋	5 個
淡忌廉	40 毫升
鹽、黑胡椒	適量
牛油	50 克
煙三文魚（切粒）	120 克
薑（去皮，切碎）	5 克
細香葱（切碎）	20 克

上 碟

煙三文魚	300 克
初榨橄欖油	10 毫升

裝 飾

豌豆苗	適量
法國芫茜	適量
細香葱	適量
初榨橄欖油	適量

做法

炒 蛋

1. 蛋、淡忌廉和鹽略攪拌至剛好混合。

2. 易潔平底鑊加熱約 1 分鐘，加入 1/3 牛油，略煎薑粒，盛起備用。

3. 同一平底鑊煮溶餘下牛油，牛油不要煮至變色，否則炒蛋會變暗不夠漂亮。

4. 倒入蛋漿，頭 5-10 秒靜置不要攪拌，再用膠鏟將蛋漿從鑊底往上覆，再靜置 5-10 秒，再攪拌和往上覆蛋漿。

5. 重複以上步驟直至蛋漿剛剛開始凝固，下薑粒、煙三文魚和細香葱。

完 成

6. 將煙三文魚放在碟上呈圓形。

7. 放上炒蛋和裝飾，灑少許初榨橄欖油即可享用。

提示

不要使用大火煮蛋漿，否則蛋會太「老」，不夠嫩滑。

Smoked salmon scrambled eggs flavoured with ginger

INGREDIENTS

SCRAMBLED EGGS

- 5 pcs fresh egg
- 40 ml whipping cream
- salt to judge
- 50 g butter (for cooking)
- 120 g smoked salmon, diced
- 5 g ginger, peeled and finely chopped
- 20 g chive, chopped

FOR ASSEMBLING

- 300 g smoked salmon, sliced
- 10 ml extra virgin olive oil

GARNISHES

- bean sprout
- chervil
- chive
- extra virgin olive oil

COOKING TIPS

- Don't use high heat while cooking the eggs, otherwise it will be overcooked and tough.

METHOD

SCRAMBLED EGGS

1. Lightly whisk the eggs, cream and a pinch of salt together until all the ingredients are just combined.

2. Heat a small non-stick frying pan for a minute or so, then add 1/3 of the butter and lightly sauté the chopped ginger, put it aside.

3. Using the same pan, melt the remaining butter. Don't allow it to brown or it will discolour the eggs.

4. Pour in the egg mixture and let it sit, without stirring, for 5-10 seconds. Stir with a spatula, lifting and folding it over from the bottom of the pan. Let it sit for another 5-10 seconds then stir and fold again.

5. Repeat until the eggs are softly set and slightly runny in places, then return the chopped ginger and add the smoked salmon and chopped chives.

TO FINISH

6. Arrange the smoked salmon slices in a round shape on a plate.

7. Top up with the scramble egg, finally put the garnishes on top and drizzle with olive oil.

主菜 | 4 人份

慢煮美國肉眼扒
伴羊肚菌忌廉汁

Oven roasted (slow-cooked style)
US Beef ribeye with morel sauce

材料

肉眼扒

美國肉眼扒（約 1 吋厚）	500 克
橄欖油	10 毫升
蒜肉（壓碎）	15 克
新鮮百里香	10 克

薯蓉

美國焗薯	2 個
全脂牛奶	100 毫升
無鹽牛油	50 克
鹽、黑胡椒	適量

羊肚菌白汁

乾羊肚菌	30 克
雞湯或水（看 p.190）	150 毫升
乾葱（切碎）	15 克
白蘭地	25 毫升
淡忌廉	200 毫升

裝飾

雜菜球（蒸熟）	50 克

提示

+ 如果沒有微波爐，可以隔水蒸熟薯仔。
+ 羊肚菌最少浸 3 次雞湯或水，以浸走泥沙。
+ 肉眼扒焗好後一定要靜置，以便能保留最多的肉汁。

做法

薯茸

1. 薯仔整個放在微波爐裏煮約 8 分鐘，切開一半取出薯肉，用薯茸夾將薯仔壓進小鍋內，用小火加熱。
2. 逐少加入牛奶和牛油，並不斷攪拌，下鹽和黑胡椒調味，將薯茸過篩。

羊肚菌白汁

3. 乾羊肚菌浸在上湯或水內半小時，瀝乾，切碎，保留上湯或水。
4. 在小鍋內，用中火炒乾葱和羊肚菌直至軟身，灒下白蘭地，加入上湯並煮至剩下 1/3 水分，下淡忌廉煮至濃稠。

肉眼扒

5. 肉眼扒用鹽和黑胡椒調味，用橄欖油、蒜頭及百里香以大火炙至兩面金黃，肉眼扒移至焗架上，放在已預熱的焗爐內，以攝氏 100-110 度焗大約 10 至 15 分鐘（這個火力和時間，能將 1 吋厚的牛扒焗至三至四成熟）。
6. 肉眼扒移離焗爐並以錫紙覆蓋，靜置大約 10-15 分鐘。
7. 將肉眼扒切片，伴薯蓉、羊肚菌白汁和雜菜球供食。

Oven roasted (slow-cooked style) US beef ribeye with morel sauce

INGREDIENTS

BEEF RIBEYE

- 500 g US beef ribeye (about 1 inch thick)
- 10 ml olive oil
- 15 g garlic, crushed
- 10 g fresh thyme

MASHED POTATO

- 2 pcs US baking potato
- 100 ml full cream milk
- 50 g unsalted butter
- salt and pepper to taste

MOREL CREAM SAUCE

- 30 g dry morels
- 150 ml chicken stock or water, see p.190
- 15 g shallot, chopped
- 25 ml brandy
- 200 ml whipping cream

GARNISH

- 50 g mixed vegetable balls (steam to cook)

COOKING TIPS

+ The potatoes can be steamed to cook if there is no microwave oven.

+ Soak the morel mushrooms at least 3 times to remove the sand as much as you can by using the same stock or water.

+ Resting is paramount important for the beef to retain the juice once it has finished cooking in the oven.

METHOD

MASHED POTATO

1. Cook the potato with skin on in a microwave oven for about 8 minutes, then cut it in half and take the mash out. Pass it through a ricer into a sauce pan over low heat.

2. Add milk and butter bit by bit to the mashed potato and stir well with a spatula, season with salt and pepper. Finally pass it through a fine sieve.

MOREL CREAM SAUCE

3. Soak the dry morels in the chicken stock or water for about half an hour, strain and chop them. Keep the stock.

4. Sauté the chopped shallot and morels in a sauce pan over medium heat until tender, deglaze the pan with brandy. Add the stock and let it reduced by 2/3 and finally add the cream. Let it reduce again until thick.

COOK THE BEEF

5. Season the beef with salt and pepper; sear it until golden brown on both sides in a hot pan with olive oil, garlic and thyme. Transfer it onto a wire rack and slow cook it in a pre-heated oven (about 100-110°C) for about 10 to 15 minutes (medium rare for 1-inch steak)

6. Then take it out from the oven and cover it with a foil, let it rest for about 10 to 15 minutes.

7. Then cut it into slices and serve them with the mashed potato, sauce and mixed vegetable balls.

慢煮美國肉眼扒伴羊肚菌忌廉汁

甜點 | 4 人份

法式冧酒蛋糕
Rum Baba

蛋糕粉漿

中筋麵粉	110 克
鹽	1 克
乾酵母	3.5 克
暖牛奶	30 毫升
蜂蜜	8 克
蛋漿（室溫）	100 克
＊無鹽牛油（室溫放軟）	55 克

＊預備多一點軟牛油，用來塗抹蛋糕模

冧酒糖漿

水	250 毫升
幼砂糖	60 克
檸檬皮	1/3 個份量
橙皮	1/3 個份量
雲尼拿豆莢	1/2 條
黑冧酒	50 毫升

淡忌廉

淡忌廉	100 毫升
糖霜	20 克
雲尼拿豆莢	1/2 條

裝飾

紅莓	8 顆
糖霜	10 克
羅勒葉	4 片

做法

蛋糕粉漿

① 將麵粉過篩進座枱攪拌機的大碗內,下鹽攪拌;用另外一個大碗將牛奶、蛋、蜂蜜和酵母拌勻。

② 攪拌機調校至中速,邊攪拌邊下蛋漿混合物。

③ 慢慢加入室溫牛油,並持續攪拌 4-5 分鐘,直至粉漿軟滑有光澤。

④ 用牛油抹勻圓形中空蛋糕模,並灑入少許麵粉;粉漿用唧袋唧進模內,直至半滿,靜置在暖的地方(約攝氏 25 度),讓粉漿發酵膨脹至兩倍大。

⑤ 放入焗爐,以攝氏 180-190 度焗 15 分鐘,直至膨脹和顏色金黃。

冧酒糖漿

⑥ 剝開雲呢拿豆莢,刮出雲尼拿籽。

⑦ 除了冧酒,將所有材料放在小鍋內,用中火加熱,不斷攪拌直至糖溶,煮至水分減少一半,待涼,加入冧酒並過篩,冷藏。

淡忌廉

⑧ 剝開雲呢拿豆莢,刮出雲尼拿籽。

⑨ 淡忌廉和糖霜放入大碗內,攪拌直至忌廉企身,加雲尼拿籽拌勻,放在唧袋內備用。

完成

⑩ 將焗好的蛋糕浸在糖漿內 4-5 秒,放在湯碗裏,上面唧忌廉,以紅莓、糖霜和羅勒葉裝飾。

提示

+ 蛋糕模要先塗抹上牛油和灑少許麵粉,否則焗後的蛋糕難以移離蛋糕模。

+ 不要讓麵糰發酵太久,讓它膨脹至模邊就可以,否則蛋糕會焗得像鬆餅般拱起。

Rum Baba

INGREDIENTS

BABA BATTER

- 110 g plain flour
- 1 g salt
- 3.5 g instant dry yeast
- 30 ml milk (warm)
- 8 g honey
- 100 g beaten eggs (room temperature)
- 55 g unsalted butter, soften at room temperature, plus extra for greasing

BABA SYRUP

- 250 ml water
- 60 g castor sugar
- 1/3 pc lemon zest
- 1/3 pc orange zest
- 1/2 pc vanilla pod
- 50 ml dark rum

FOR THE WHIPPED CREAM

- 100 ml whipping cream
- 20 g icing sugar
- 1/2 pc vanilla pod

GARNISH

- 8 pcs raspberry
- 10 g icing sugar
- 4 pcs basil leaf

BABA BATTER

① Sift the flour into a large bowl of a free stand mixer fitted with a petal and stir in the salt. Whisk the milk, eggs, honey and yeast in a separate bowl until well combined.

② Gradually add the egg mixture into the flour mixture to create a smooth batter while the machine is running at medium speed.

③ Slowly add the softened butter into the batter. Beat the mixture for about 4-5 minutes or until completely smooth and glossy.

④ Transfer the batter into a piping bag and pipe it into the greased and floured moulds (Sararin moulds) until they are half full and set aside in a warm place (about 25°C) to rise to roughly double in size.

⑤ Bake the baba in the pre-heated oven (180-190°C) for 15 minutes until risen and golden brown.

BABA SYRUP

⑥ Split open the vanilla pod and scrape out the seeds.

⑦ Mix everything together except rum in a sauce pan over medium heat. Stir constantly until all sugar has dissolved, let it reduce by half. Let it cool, then add the rum and strain it through a fine sieve. Keep chilled.

FOR THE WHIPPED CREAM

⑧ Split open the vanilla pod and scrape out the seeds.

⑨ Combine the cream and icing sugar together in a mixing bowl, beat until stiff and add the vanilla seeds and stir well. Transfer it into a piping bag and ready for later use.

TO FINISH

⑩ Soak the baba cake in syrup for about 4-5 seconds and place them on a soup bowl. Pipe the whipped cream on top and assemble the raspberries on and garnish with icing sugar and basil leaves.

COOKING TIPS

+ Make sure the moulds are well greased and floured before piping in the batter otherwise it is difficult to remove the cake out.

+ Don't over prove the batter (just let the batter expand almost to the top) otherwise you will get a muffin top around the edges.

法式冧酒蛋糕

151

SET/02

winter menu

Starter
頭盤

煎 蟹 餅
配 榛 子 蘋 果 茴 香 沙 律

Crab cake
with hazelnut, apple and fennel salad

構 思 HOW IS IT DESIGNED ?

這個套餐相比前一個套餐，需要有更多煮食技巧，尤其是捲起雞胸肉的步驟。不過，有些步驟可以預先做好的，能縮短煮食及上菜的時間。至於味道方面，帶有檸檬香的忌廉醬，配上雞胸、芝士和火腿的滋味一流。最後，蘋果撻的甜味可以中和主菜濃郁的味道。

This menu is more difficult than the first one, it takes a little more skills, especially how to roll up the chicken breast, but some of the steps can be prepared in advance so as to shorten the cooking time and make the service smoother. Taste wise, the lemony cream sauce is perfectly matched with the chicken breast cooked with cheese and ham. Lastly, the sweetness from the apple tart tatin can cut off the heaviness from the main course.

Main Dish

主菜

Dessert

甜點

芝士雞卷
伴牛油甘筍

Chicken Cordon Bleu
with glazed carrot

蘋果撻
配冧酒提子雪糕

Apple tarte tatin
with rum and raisin ice cream

預先準備 TIPS FOR SMART PREPARATION

+ 蟹餅和雞卷可以預早一天捲起造型，放在雪櫃冷藏。

+ 雪糕可以預先做好，放在冰格冷凍。

+ The crab cake and chicken rolls can be shaped and rolled up one day before, then keep them in the fridge to set.

+ The ice cream can be made in advance and keep in the freezer.

頭盤 | 4 人份

煎蟹餅
配榛子蘋果茴香沙律

Crab cake with hazelnut, apple and fennel salad

材料

蟹 餅

洋葱（切碎）	50 克
紅、黃西椒（切粒）	80 克
橄欖油	30 毫升
罐頭蟹肉	250 克
麵粉	10 克
檸檬汁及皮	1/3 個
鹽、黑胡椒	適量

沙 律 汁

焗香榛子（壓碎）	50 克
檸檬汁	5-10 毫升
初榨橄欖油	50 毫升
鹽、黑胡椒	適量

沙 律

茴香（切薄片）	1 個
青蘋果（切條）	半個
沙律生菜	50 克

做法

蟹 餅

1. 用橄欖油炒洋葱和西椒直至十分乾身，備用。

2. 瀝乾蟹肉，用廚房紙抹乾；所有材料放在大碗內拌勻，將混合物放在圓模內造成蟹餅，冷藏。

沙 律 汁

3. 所有材料放在大碗內拌勻，備用。

上 菜

4. 用橄欖油煎蟹餅直至兩面金黃和內裏和暖，上碟。將沙律和沙律汁拌勻，放在蟹餅上，再澆些沙律汁，即可享用。

提示

＋ 建議在煎蟹餅前，兩面撲上麵粉，蟹餅在煎熟後表皮會很鬆脆。

STARTER | FOR 4 PORTIONS

Crab cake with hazelnut, apple and fennel salad

INGREDIENTS

CRAB CAKE

- 50 g onion, chopped
- 80 g yellow and red bell pepper, diced
- 30 ml olive oil
- 250 g crab meat, tinned
- 10 g plain flour
- 1/3 pc lemon juice and zest
- salt and pepper to taste

FOR THE DRESSING

- 50 g hazelnuts, toasted and crushed
- 5-10 ml lemon juice
- 50 ml extra virgin olive oil
- salt and pepper to taste

FOR THE SALAD MIX

- 1 bulb fennel, thinly sliced
- 1/2 pc green smith apple, cut into strips
- 50 g frisée lettuce

METHOD

CRAB CAKE

1. Sweat the chopped onions and peppers with olive oil until very dry, keep aside.

2. Strain the crab meat and pat it dry with kitchen paper. Mix it with all the ingredients in a mixing bowl, mould the mixture into small cakes. Keep chilled to set.

FOR THE DRESSING

3. Mix everything together in a mixing bowl, stir to mix. Keep it aside.

TO FINISH

4. Pan fry the crab cake with olive oil until golden brown on both sides and warm inside. Toss the salad mix with the dressing and arrange them on the cakes. Drizzle with extra dressing around and serve.

COOKING TIPS

+ Dust the crab cake on both sides with extra flour before pan frying so as to get crispier skin.

主菜 | 4 人份

芝士雞卷
伴牛油甘筍

Chicken Cordon Bleu
with glazed carrot

材料

雞 卷

無皮雞胸	2 件
新鮮百里香葉	10 克
巴馬火腿或意大利火腿	4 片
水牛芝士碎（Mozzarella）	25 克
瑞士硬芝士碎（Gruyere）	25 克
鹽、黑胡椒	適量

漿 粉

麵粉	20 克
全蛋（打勻）	1 個
麵包糠	100 克

牛 油 甘 筍

甘筍	400 克
乾葱（切碎）	50 克
無鹽牛油	50 克
幼砂糖	20 克
水	100 毫升
新鮮橙汁	50 毫升
意大利芫茜（切碎）	5 克
鹽、黑胡椒	適量

裝 飾

法國芫茜（chervil）	5 克

提示

＋ 可以將雞卷放在冰格冷凍 1 小時，加快冷卻速度。

做法

雞 卷

1. 修齊雞胸不規則的邊，再切雙飛，打拍成 0.5 厘米厚的長方形。

2. 雞胸用百里香、鹽和黑胡椒調味，放上兩片火腿，下芝士碎，用保鮮紙將雞胸捲起，放在雪櫃冷藏。餘下的雞胸，同樣做法。

3. 移去保鮮紙，雞卷先拍上麵粉，蘸蛋汁，再裹上一層麵包糠，放在雪櫃冷藏。

牛 油 甘 筍

4. 甘筍切薄片；用小鍋中火熱溶牛油，下乾葱，炒至乾葱透明，加入甘筍片拌勻，下鹽和黑胡椒調味。

5. 灑入鹽和糖，拌勻，再炒約 2-3 分鐘，下水和橙汁，加蓋煮約 3-5 分鐘，直至甘筍變軟。

6. 開蓋，並轉至中大火，煮至汁液乾身及甘筍有光澤，下意大利芫茜，拌勻。

完 成

7. 用橄欖油和牛油半煎炸雞卷，直至表皮金黃，放入已預熱攝氏 140-150 度焗爐焗約 15 分鐘。

8. 將雞卷切件，伴以牛油甘筍和法國芫茜享用。

Chicken Cordon Bleu with glazed carrot

INGREDIENTS

FOR THE CHICKEN

- 2 pcs chicken breast skinless
- 10 g fresh thyme leaves
- 4 pcs Parma ham or prosciutto
- 25 g mozzarella, grated
- 25 g gruyere cheese, grated
- salt and pepper to judge

FOR THE COATING

- 20 g plain flour
- 1 pc whole egg
- 100 g breadcrumbs

BUTTER GLAZED CARROT

- 400 g carrot
- 50 g shallot, chopped
- 50 g unsalted butter
- 20 g castor sugar
- 100 ml water
- 50 ml fresh orange juice
- 5 g Italian parsley, chopped
- salt and pepper to taste

GARNISH

- 5 g chervil

COOKING TIPS

+ Put the chicken rolls into the freezer for about an hour to speed up the chilling process.

METHOD

FOR THE CHICKEN

1. Trim the chicken breasts and butterfly them, pound them into about 0.5 cm thick rectangular shape.
2. Season the chicken breast with thyme, salt and pepper. Place two slices of ham on top, and sprinkle the cheese on. Roll up each breast by using cling film. Let them set in the fridge.
3. Remove the cling film and thinly coat them with flour, then dip into the beaten eggs. Finally coat them evenly with breadcrumbs. Keep them in the fridge for service.

BUTTER GLAZED CARROT

4. Slice the carrots into thin slices evenly. Melt the butter in a sauce pan over medium heat, add the chopped shallot and sweat them until transparent. Then add the carrot and toss to combine. Season with salt and pepper.
5. Sprinkle the salt and sugar over carrots and toss again. Sauté for 2 to 3 minutes, then add water and orange juice. Cover the pan and cook for another 3-5 minutes or until they are tender.
6. Uncover the pan and increase heat to medium-high and cook away most of the liquid until the carrots coat on a layer of shiny butter. Finally add the chopped parsley.

TO FINISH

7. Shallow fry the chicken breast with a mixture of olive oil and butter until golden brown on all sides, finish cooking at the pre-heated oven at (140-150°C) for about 15 minutes.
8. Cut the chicken into slices and serve with the glazed carrot. Finally garnish with chervils.

芝士雞卷伴牛油甘筍

159

甜點 | 4 人份

蘋果撻
配冧酒提子雪糕

Apple tarte tatin
with rum and raisin
ice cream

材料

蘋果餡

加拿蘋果	3 個
幼砂糖	100 克
水	40 毫升
無鹽牛油（切粒）	50 克
黑冧酒	100 毫升
雲尼拿豆莢（刮出籽）	1 條

其他

酥皮（已壓平）	200 克
蛋黃（掃酥皮用）	20 克
* 冧酒提子雪糕	4 球
* 做法看 P137：棗子拖肥蛋糕 （秋天）	

裝飾

糖霜	10 克

做法

煮蘋果

① 蘋果去皮切半，用茶匙刮走蘋果心。

② 小鍋內下幼砂糖和水，用大火加熱，煮至開始變成淡黃色焦糖，移離爐火，下牛油，並搖擺小鍋至牛油溶化，下冧酒和雲尼拿籽。

③ 下蘋果煮 5-6 分鐘，期間需要不斷反轉蘋果直至變軟，靜置冷卻。

完成

④ 將蘋果放入多個細蛋糕模內，每個模內最少下 1 湯匙糖漿。

⑤ 酥皮切成蛋糕模大小，並蓋着全部蛋糕模，將邊緣的酥皮按下，蛋黃塗在酥皮上。

⑥ 放入已預熱焗爐，以攝氏 180-190 度焗 20-25 分鐘，直至酥皮金黃。

⑦ 小心將蘋果撻轉至湯碟內，灑下糖霜，伴冧酒提子雪糕享用。

提示

+ 切掉的酥皮可以焗熟作裝飾。

+ 煮好的蘋果要徹底冷卻，否則蓋上酥皮時，酥皮遇熱會變軟。

Apple tarte tatin with rum and raisin ice cream

INGREDIENTS

APPLE FILLING

- 3 pcs Gala apple
- 100 g castor sugar
- 40 ml water
- 50 g unsalted butter, cubed
- 100 ml dark rum
- 1 pc vanilla pod, seeds scraped out

OTHERS

- 200 g puff pastry, already rolled
- 20 g egg yolk, for brushing the pastry
- *4 scoops rum and raisin ice cream
  *Refer to P.137 rum and raisin ice cream, autumn Sticky toffee pudding with dates.

GARNISH

- 10 g icing sugar

TO COOK THE APPLES

① Peel the apples, then halve them horizontally and use a teaspoon to get rid of the core.

② Combine sugar and water together in a saucepan over high heat, cook the mixture until it forms a light caramel. Remove it from the heat and add the butter, shake the pan to melt. Then add the rum and vanilla seeds.

③ Add the halved apples. Cook, keep turning the apples constantly, for about 5-6 minutes or until the apples start to soften. Let it cool completely.

TO FINISH

④ Arrange the apples into the ramekin dishes, spoon at least 1 table of cooking juice into each dish.

⑤ Cover it quickly with the round pre-cut puff pastry sheets. Carefully tuck the pastry down right into the edges. Brush the pastry with beaten egg yolk.

⑥ Bake the tarte tartin in the pre-heated oven at 180-190°C for about 20 to 25 minutes, or the pastry is until golden.

⑦ Turn the tarte tartin carefully onto a soup plate and serve it with the ice cream and icing sugar.

COOKING TIPS

+ Keep the scraps of puff pastry and bake them as well for garnish.

+ Chill the cooked apples completely so that the pastry won't get too soft when put on.

蘋果撻配冧酒提子雪糕

163

SET/03

winter
menu

栗 子
青 蘋 果 湯

Winter chestnut
and green apple soup

構 思　HOW IS IT DESIGNED ?

這個套餐是向難度挑戰，做法比之前的複雜，需要更加多技巧，尤其是主菜。雖然烹調主菜是很花時間，但是當你上菜招呼朋友時，看見友人品嘗這個矜貴又美味的鴨肝脆批時流露的滿足，感覺是非常值得的。相比主菜，頭盤和甜點的準備工夫和烹調過程比較簡單，尤其是甜點，讓你有更多時間準備主菜。

This menu is even more challenging when compared with the previous one particularly the main course. It is time consuming but it is worthwhile at the end if they are being served individually on the table in front of your close friends. It looks really luxurious and tastes marvelous. Compared with the main course, the starter and dessert are much easier to prepare, especially the dessert, saves you much time to compensate for the time spend on the main course.

Main Dish

主菜

Dessert

甜點

鴨 肝 脆 批
配 松 露 甜 酒 汁

Duck foie gras pithivier
with truffle and madeira sauce

法 式
橙 酒 班 戟

Crepe Suzette

預 先 準 備 TIPS FOR SMART PREPARATION

+ 鴨肝脆批的餡料可以預早一天做好。

+ 松露甜酒汁也可以預早做好。

+ 法式班戟可以當天早上做好，夜晚
才奉客。

+ 建議使用市售的高質素即用酥皮，可
以縮短煮食時間。

+ All the fillings for making pithivier can be
finished one day before.

+ The Madeira sauce can also be finished in
advance.

+ The crepes can be finished in the morning
time and used for service during the night.

+ Use the good quality already-made puff
pastry sheet to shorten the preparation
time.

頭盤 | 4 人份

栗子青蘋果湯
Chestnut and green apple soup

材料

湯

栗子肉（切片）	250 克
無鹽牛油	40 克
青蘋果（切粒）	半個
西芹（切片）	50 克
蒜頭（切片）	5 克
百里香葉	2 棵
雞湯（看 p.190）	250 毫升
全脂牛奶	250 毫升
鹽、黑胡椒	適量

配料

栗子肉（切片）	10 克
青蘋果（切條）	半個
帶子（刺身級別）	4 隻

做法

湯

1. 用小鍋及中火煮溶牛油，下栗子肉煮約 10 分鐘，加入青蘋果、西芹、蒜頭和百里香，邊煮邊攪拌煮多 10 分鐘。

2. 倒入雞湯和牛奶，煮滾後轉小火煮 30-40 分鐘，直至栗子肉變軟，下鹽和黑胡椒調味。

3. 然後倒入攪拌機內，打成軟滑栗子青蘋果湯，過篩。

完成

4. 翻熱湯並倒入湯碗內，以栗子肉、青蘋果和煎帶子伴食。

提示

+ 建議使用兩種栗子（已剝皮及帶殼的），煮好的湯會特別美味。
+ 煲湯時會流失水分，可酌量加水。
+ 可以用打蛋器攪拌湯，以製作泡沫。

Chestnut and green apple soup

INGREDIENTS

SOUP

- 250 g fresh chestnuts, cut into slices
- 40 g unsalted butter
- 1/2 pc green apple, cut into dices
- 50 g celery, cut into slices
- 5 g garlic, sliced
- 2 sprigs thyme leaves
- 250 ml chicken stock, see p.190
- 250 ml full cream milk
- salt and pepper to taste

GARNISHES

- 10 g chestnut slices
- 1/2 pc green apple, cut into strips
- 4 pcs scallop, sashimi grade

METHOD

FOR THE SOUP

1. Melt the butter in a sauce pan over medium heat, add the sliced chestnut and cook for about 10 minutes. Add the green apples, celery, garlic slices and thyme; continue cooking, keep stirring constantly, for another 10 minutes.

2. Pour in the chicken stock and milk, bring them to a boil and turn the heat down to simmer for about 30 to 40 minutes until the chestnut is tender. Season with salt and pepper.

3. Transfer the mixture into a blender and blend it until smooth, then pass it through a fine sieve.

TO FINISH

4. Reheat the soup and pour it into the soup bowls and garnish with freshly sliced chestnut, apple juliennes and pan-fried scallops.

COOKING TIPS

- Try to use two types of chestnut (peeled one and the one with shells) to give a special flavour to the soup.
- Add water if necessary to replenish the loss of cooking liquor during cooking.
- Use a hand blender to blend the soup lightly to make foam.

主菜 | 4 人份

鴨肝脆批配松露甜酒汁

Duck foie gras pithivier
with truffle and Madeira sauce

B

材料

材 料

巴馬火腿	2 片（切半）
肥鴨肝（煎好冷卻）	160 克
大椰菜葉（焓熟）	2 塊
酥皮	4 張
蛋黃（掃酥皮用）	30 克

蘑 菇 醬

大燒烤蘑菇	4 個
乾蔥（切片）	20 克
蒜頭（切片）	20 克
牛油	20 克
鹽、黑胡椒	適量

焦 糖 洋 蔥

煙燻肉（切粒）	50 克
白洋蔥（切片）	2 個
無鹽牛油	50 克
蒜頭（切片）	20 克
百里香	5 克
意大利黑醋	10 毫升
鹽、黑胡椒	適量

黑 松 露 甜 酒 汁

雞翼（切件）	400 克
蒜頭（壓碎）	20 克
橄欖油	10 毫升
洋蔥（切粒）	100 克
葡萄牙馬德拉酒（Madeira）	100 毫升
雞湯（看 p.190）	300 毫升
牛油麵粉糊（看 p.195）	20 克
松露醬、松露油	適量
鹽、黑胡椒	適量

A

提示

+ 放椰菜在加了鹽的滾水焓 30 秒，撈起放入冰水內冷卻，瀝乾。

+ 椰菜球可以用保鮮紙輔助做成圓形。（圖 AB）

+ 最少塗兩次蛋黃在酥皮上，令批皮焗好後更加金黃漂亮。

蘑菇醬

❶ 大蘑菇切粒。用平底鑊及中火將牛油熱溶，倒入蘑菇粒、乾葱和蒜頭炒至水分蒸發，下鹽和黑胡椒調味。

❷ 倒入食物處理器內打成幼粒，冷藏。

焦糖洋葱

❸ 用小鍋將煙肉煎香至出油，下洋葱、蒜頭和牛油，煎至洋葱透明並焦糖化，加入百里香、意大利黑醋，下鹽和黑胡椒調味，待涼。

黑松露甜酒汁

❹ 用中大火燒熱平底鑊，加入橄欖油和蒜頭，將雞翼煎至表面金黃，下洋葱，轉至小火，將洋葱煮至變軟。

❺ 注入馬德拉酒，煮至汁液收至三分一，下雞湯煮 30 分鐘。將汁液過篩，下牛油麵粉糊煮至濃稠，加入松露醬和油，下鹽和黑胡椒調味。

組合

❻ 放一個 6 厘米圓形模在墊上，舀入焦糖洋葱，壓實，移開圓形模。將蘑菇醬放在洋葱上面，放一件鴨肝及火腿，用椰菜葉包起所有材料，裹成圓形。

⑦

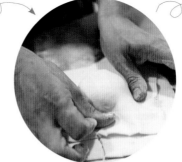

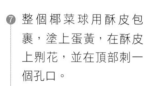

⑦ 整個椰菜球用酥皮包裹，塗上蛋黃，在酥皮上剶花，並在頂部刺一個孔口。

⑧ 放入已預熱攝氏 200 度的焗爐內，焗約 10 分鐘至金黃，轉至攝氏 160 度再焗 5-8 分鐘，直至所有餡料熱透。

完 成

⑨ 將鴨肝脆批放在碟中間，伴黑松露甜酒汁享用。

Duck foie gras pithivier
with truffle and Madeira sauce

INGREDIENTS

INGREDIENTS

- 2 pcs Parma ham, cut into halves
- 160 g duck foie gras, pan fried and cooled
- 2 pcs cabbage leaves, blanched (big)
- 4 sheets puff pastry sheet
- 30 g egg yolk (for brushing the pastry)

MUSHROOM DUXELLES

- 4 pcs portobello mushrooms
- 20 g shallot, sliced
- 20 g garlic, sliced
- 20 g butter
- salt and pepper to taste

CARAMELIZED ONIONS

- 50 g smoked bacon, diced
- 2 pcs white onion, sliced
- 50 g unsalted butter
- 20 g garlic, sliced
- 5 g thyme
- 10 ml balsamic vinegar
- salt and pepper to taste

TRUFFLE JUS

- 400 g chicken wings, cut into pieces
- 20 g garlic, crushed
- 10 ml olive oil
- 100 g onion, cut into cubes
- 100 ml Madeira wine
- 300 ml chicken stock, see p.190
- 20 g beurre manie, see p.195
- truffle paste and oil to taste
- salt and pepper to taste

鴨肝脆批配松露甜酒汁

173

MUSHROOM DUXELLES

1. Cut the mushrooms into dices and sauté them with butter, shallot and garlic in a pan over medium heat until all water has evaporated. Season to taste with salt and pepper.

2. Transfer them into a processor and process until coarse. Keep chilled.

CARAMELIZED ONIONS

3. Render the bacon in a sauce pan, then add the sliced onions, garlic and butter, sweat them until soft and nicely caramelized. Finally add the thyme, balsamic vinegar, season to taste with salt and pepper. Leave it to cool.

MADEIRA AND TRUFFLE SAUCE (JUS)

4. Sauté the chicken wings with olive oil and garlic in a frying pan over medium high heat until golden brown, then add onions and continue cooking over low heat until they are soft.

5. Add Madeira and reduce by 2/3, then add the chicken stock and let it simmer for about 30 minutes. Strain the sauce and thicken it by beurre manie, finally add the truffle paste and oil. Season to taste with salt and pepper.

TO ASSEMBLE

6. Use a 6-cm diameter ring sitting on a mat, press the caramelized onion at the bottom and top up with the mushroom duxelles. Put the one piece of foie gras and ham on top, finally wrap the whole thing with blanched cabbage leaves. Make sure they are in nice round shape.

7. Finally cover the cabbage ball with puff pastry. Brush them with egg yolk, score on the surface and prick a hole on top.

8. Bake them in the pre-heated oven at 200°C until golden brown (around 10 minutes), then turn the heat down to 160°C, bake it for another 5-8 minutes or until it hot inside.

TO FINISH

9. Place the pithivier at the middle of a plate, drizzle the sauce around and serve immediately.

COOKING TIPS

+ Blanch the cabbage leaves in salted boiling water for about 30 seconds and stop cooking by plunging into ice water, strain and dry.

+ The cabbage ball can be shaped rounder by using cling film.

+ Brush the pithivier with egg yolk at least 2 times to get a better golden brown colour.

甜點 | 4 人份

法式橙酒班戟
Crepe Suzette

材料

班戟

麵粉	110 克
蛋	100 克約 2 個
牛奶	275 毫升
無鹽牛油（已溶）	30+20 克
橙皮茸	半個份量
幼砂糖	15 克
鹽	少許

橙酒汁

幼砂糖	20 克
無鹽牛油	50 克
橙汁	150 毫升
橙皮	半個
檸檬汁和皮	半個
柑曼怡／君度橙酒	50 毫升
新鮮橙（切件）	1 個

其他

新鮮薄荷	4 棵
淡忌廉	50 毫升 ⎤ 打起
雲尼拿籽	半條份量 ⎦

做法

班戟

1. 麵粉和鹽一起篩在大碗內，蛋打成蛋漿，蛋漿分三次加入麵粉內，拌勻，邊攪拌邊慢慢加入牛奶，攪至沒有粉粒，過篩。
2. 拌入橙皮茸、糖和 30 克溶牛油，其餘 20 克牛油抹平底鑊用。
3. 在平底鑊內煎班戟，將班戟疊起，備用。

橙酒汁

4. 用不鏽鋼平底鑊及大火煮溶幼砂糖成淺色焦糖，下牛油煮至溶，加入橙汁和檸檬汁。
5. 煮至汁液收乾至一半或開始濃稠，下其他材料拌勻。

完成

6. 班戟放入橙酒汁內，慢慢加熱，摺成三角形並鋪在碟上。
7. 倒入橙酒，並用火槍燒（非必要）。
8. 班戟伴以橙酒汁、暖橙件和打起的雲尼拿忌廉、綴上薄荷裝飾。

提示

+ 煮橙酒汁時，要將糖逐少灑在熱平底鑊內，否則會很快煮焦。
+ 用火槍燒橙酒的步驟並非必需，可以不做。

Crepe Suzette

INGREDIENTS

FOR THE CREPES

- 110 g plain flour
- 100 g (2 pcs) whole eggs
- 275 ml milk
- 30+20 g unsalted butter, melted
- 1/2 pc orange zest
- 15 g castor sugar
- pinch salt

FOR THE SAUCE

- 20 g castor sugar
- 50 unsalted butter
- 150 ml orange juice
- 1/2 pc orange, zest only
- 1/2 pc lemon zest and juice
- 50 ml Grand Marnier/Cointreau
- 1 pc fresh orange segments

OTHERS

- 4 sprigs fresh mint
- 50 ml whipping cream
- 1/2 pc vanilla pod, seeds scraped out ← whipped

COOKING TIPS

+ While making the sauce, sprinkle the sugar bit by bit on the hot pan otherwise it will get burnt so easily.

+ Don't light on the flame if you are not confident of handling.

METHOD

FOR THE CREPES

1. Sift the flour and salt into a mixing bowl, whisk the eggs and add into the sieved flour in 3 times. Then gradually add small quantities of milk, stir constantly to prevent any lumps formed. Strain afterwards.

2. Finally stir in the orange zest, sugar and 30 g of melted butter, and keep the rest to lubricate the pan.

3. Cook the crepes in an oiled non-stick pan, stack together on a plate for service.

FOR THE SAUCE

4. Make a light caramel by melting the sugar on a stainless steel pan over high heat, add the butter and shake to melt. Add the orange and lemon juice.

5. Let it reduce by half or until it is slightly thick, add the remaining ingredients.

TO FINISH

6. Add the crepes to the sauce and allow them to heat gently. Then fold them into triangle shape and arrange on a plate.

7. Now you can flame them by adding extra liqueur (optional step).

8. Serve them with the sauce, warm orange segments and the vanilla whipped cream on top. Finally garnish with mint.

法式橙酒班戟

177

SET/04
winter
menu

意 式 羅 勒 青 醬
雜 菜 湯

Minestrone
flavoured with pesto

構 思 HOW IS IT DESIGNED ?

這個素食套餐有意大利的風味。
意式雜菜湯非常健康和平實，你可以加入任何蔬菜；加添了羅勒青醬的雜菜湯，為味道添加了不同層次，讓你彷彿處身於意大利鄉村的餐桌旁。薯仔糰容易製作，它的樸實滋味與南瓜的甜味互相平衡。黑朱古力梳乎厘跟一般的梳乎厘的質感有所不同，讓你在寒夜裏感到滿足和溫暖。

This menu is kind of Italian style, minestrone is a healthy and rustic soup where you can add whichever vegetables like; but here the pesto has added a twist to the soup. Gnocchi is simple to make, but it is balanced quite well by the sweetness and nuttiness from the butternut squash. The chocolate soufflé, is quite different from the normal one in terms of texture, gives you warm and full satisfaction in a cold night.

主菜

甜點

意 式 煎 薯 仔 糰
伴 鼠 尾 草 南 瓜 蓉

Pan fried potato Gnocchi
with butternut squash and sage

黑 朱 古 力
梳 乎 厘

Rich dark chocolate soufflé

預 先 準 備 TIPS FOR SMART PREPARATION

+ 羅勒青醬可以預先做好，鋪在一個墊了烘焙紙的盤上，然後放入冰格雪硬，切成一件件，儲存在密封容器內，再放入冰格，日後取出使用。

+ 薯仔糰可以預早一天烚好，到當天再煎至金黃享用。

+ 梳乎厘的粉漿可以預早一天做好，在宴客時才拌入打起蛋白再烘焙。

+ The pesto sauce can be pre-made, and then pour over a flat tray lined with parchment paper. Freeze it until hard, and then cut it into the required sizes and keep them in an air-tight container for future use.

+ The gnocchi can be finished one day before and pan fry until golden brown during the service.

+ The soufflé base can be prepared in advance or even one day before, just mix it with the meringue (whipped egg white) during the service.

🦢 頭盤 | 4人份

意式羅勒青醬雜菜湯

Minestrone flavoured with pesto

做法

雜 菜 湯

甘筍（切粒）　　　　　　50 克
洋葱（切粒）　　　　　　50 克
西芹（切粒）　　　　　　30 克
黃色意大利青瓜（切粒）　30 克
青色意大利青瓜（切粒）　30 克
紅西椒（切粒）　　　　　20 克
粟米芯（切粒）　　　　　30 克
番茄（去皮，去籽，切粒）20 克
蘆筍（切粒，焓熟）　　　50 克
四季豆（切粒，焓熟）　　50 克
雞湯（看 p.190）　　200 毫升
水　　　　　　　　　400 毫升
鹽、黑胡椒　　　　　　　適量

羅 勒 青 醬

新鮮羅勒葉　　　　　　　35 克
焗香松子（冷卻）　　　　25 克
巴馬臣芝士　　　　　　　10 克
橄欖油　　　　　　　　50 毫升
蒜頭（切碎）　　　　　　5 克
鹽、黑胡椒　　　　　　　1 克

雜 菜 湯

① 將所有蔬菜切成大小相同的小粒。鍋內下橄欖油，炒甘筍、洋葱和西芹約 4-5 分鐘，倒入其他材料（除了番茄、蘆筍和四季豆）煮 5 分鐘，下鹽和黑胡椒調味。

② 注入雞湯和水煮至微滾，再煮約 15 分鐘直至所有蔬菜變軟，試味，如有需要可以再調味。

羅 勒 青 醬

③ 洗淨羅勒葉，瀝乾水分後再抹乾。

④ 松子焗至黃金色，待涼，。

⑤ 將所有材料放在攪拌機或食物處理器內，打成青醬。

完 成

⑥ 雜菜湯翻熱，下番茄、蘆筍、四季豆和羅勒青醬，煮約 1 分鐘即成。

提示

＋ 雜菜湯最美味的食法是預早一天煮好，讓雜菜浸在湯內一整夜，翌日享用時湯味更濃郁。

＋ 在翻熱湯時才加入番茄、蘆筍及四季豆，可以讓湯帶有漂亮的鮮紅和翠綠。

＋ 羅勒青醬應該在進食前才加入，否則顏色和味道會褪色。

STARTER | FOR 4 PORTIONS

Minestrone flavoured with pesto

INGREDIENTS

SOUP

- 50 g carrot, diced
- 50 g onion, diced
- 30 g celery, diced
- 30 g/each yellow and green zucchini, diced
- 20 g red bell peppers, diced
- 30 g baby corns, diced
- 20 g tomatoes, peeled, deseeded and diced
- 50 g/ each asparagus and green beans, diced and blanched
- 200 ml chicken stock, see p.190
- 400 ml water
- salt and pepper to taste

PESTO

- 35 g fresh basil leaves
- 25 g pine nuts, roasted and cool
- 10 g parmesan cheese
- 5 g garlic, chopped
- 50 ml olive oil
- 1 g salt and pepper

METHOD

FOR THE SOUP

1. Cut all the vegetables into even dices. Sweat the carrot, onion and celery in a pot with olive oil for about 4-5 minutes. Add the remaining ingredients except tomatoes, asparagus and beans, continue cooking for another 5 minutes. Season with salt and pepper.
2. Add the chicken stock and water, bring it to a simmer and cook it for about 15 minutes or until all the vegetables are tender. Adjusting the seasoning if necessary.

FOR THE PESTO

3. Wash basil leaves, strain and pat dry.
4. Roasted pine nuts until golden brown and chilled.
5. Mix all the ingredients together in a processor or blender, process them into puree.

TO FINISH

6. Reheat the soup, add the diced tomatoes, blanched asparagus and beans and pesto. Cook it for about 1 minute. Serve.

COOKING TIPS

+ Ideally the soup can be finished one day before and let the mixed vegetables macerate, and the soup would definitely taste much better the next day.
+ Add the diced tomatoes, blanched asparagus and beans at the last minute so as to give a bright red and green colour to the soup.
+ Add the pesto to the soup just before serving it otherwise its flavour and colour would lose.

主菜 | 4 人份

意式煎薯仔糰
伴鼠尾草南瓜蓉

Pan fried potato Gnocchi
with butternut squash and sage

材料

薯仔糰

美國焗薯	2 個約 500 克
麵粉	90 克
巴馬臣芝士	30 克
鹽	8 克
鼠尾草（切碎）	10 克
蛋黃	20 克
黑胡椒	少許

南瓜蓉

冬南瓜（去皮，切片）	100 克
牛油	20 克
牛奶	100 毫升
鹽、黑胡椒	適量

牛油鼠尾草醬

新鮮鼠尾草葉（切碎）	15 克
蒜頭（切碎）	10 克
牛油	80 克
焗香松子（冷卻）	30 克
鹽、黑胡椒	適量

其他

巴馬臣芝士片	40 克
煎熟冬南瓜粒	100 克
焗脆鼠尾草（看 p.196）	16 塊

Pan fried potato Gnocchi with butternut squash and sage

做法

薯仔糰

① 薯仔連皮蒸約 30 分鐘，或直至變軟，去皮。將薯仔放入薯茸夾內夾成蓉，冷卻至乾身。

② 將乾材料放入大碗內拌勻，下薯蓉，並用手拌勻，加入蛋黃、鹽、黑胡椒拌勻，再搓成不黏手、軟身的麵糰。

③ 將麵糰揉成小薯仔糰，用滾水炶熟。

④ 盛起薯仔糰，放入凍水內冷卻，瀝乾備用。

南瓜蓉

⑤ 用小鍋及中火，下牛油煎南瓜片直至微黃，加入牛奶，並收細火煮約 10-15 分鐘，直至南瓜變軟及開始乾身。

⑥ 倒入攪拌機內打成南瓜蓉。

完成

⑦ 薯仔糰煎至兩面金黃，保溫。

⑧ 用牛油煎蒜頭碎至金黃，關火，下鼠尾草和松子拌勻，下鹽和黑胡椒調味。

⑨ 薯仔糰放在碟上，舀入南瓜蓉，旁邊放芝士、煎南瓜粒和脆鼠尾草，澆下牛油鼠尾草醬供食。

提示

+ 薯仔糰煮熟後會浮面，要立刻撈起再放入凍水冷卻。

+ 薯仔糰煮好後可以拌入少許橄欖油，以防止薯粒黏在一起。

+ 鼠尾草可以放在包有保鮮紙的碟上，每次加熱一分鐘，直至變脆。

INGREDIENTS

GNOCCHI

- 500 g (2 pcs) US baking potato
- 90 g plain flour
- 30 g parmesan cheese
- 8 g salt
- 10 g sage, chopped
- 20 g egg yolk
- pinch ground black pepper

BUTTERNUT SQUASH PUREE

- 100 g butternut squash, peeled and sliced
- 20 g butter
- 100 ml milk
- salt and pepper to taste

SAGE BUTTER SAUCE

- 15 g fresh sage leaves, chopped
- 10 g garlic, chopped
- 80 g butter
- 30 g pine nuts, roasted and cool
- salt and pepper to taste

OTHERS

- 40 g parmesan cheese shavings
- 100 g butternut squash dices, sautéed
- 16 pcs crispy sage, see p.196

意式煎薯仔糰伴鼠尾草南瓜蓉

METHOD

GNOCCHI

1. Steam the potato with the skin for about 30 minutes or until soft, then peel. Scoop out the potato and pass it through a potato ricer, let it cool and dry.

2. Mix all the dried ingredients in a mixing bowl, add the cool potato and mix well by hands. Finally add egg yolk, salt and pepper, knead lightly to form a soft and non-sticky dough.

3. Roll the dough into desired shapes and cook them in boiling water until cooked.

4. Remove and cool them down in cold water, strain for later pan frying.

BUTTERNUT SQUASH PUREE

5. Sauté sliced squash with butter in a sauce pan over medium heat until lightly brown. Add the milk and turn the heat down and let it cook for about 10-15 minutes or until it is tender and almost dry.

6. Transfer the mixture into blender and blend into a fine puree.

TO FINISH

7. Pan fry the gnocchi until golden brown on both sides, keep warm.

8. Melt the butter with the chopped garlic on the same pan until nutty brown, let it cool down a bit and add the chopped sage and toasted pine nuts. Stir to mix.

9. Arrange the squash puree and gnocchi onto a plate, place the cheese, sautéed squash dices and crispy sage on the sides. Finally drizzle the sauce around.

COOKING TIPS

+ The gnocchi will float on top when they are cooked. Remove them straight away and Cool Down in cold water.

+ Add a bit of olive oil to the gnocchi to prevent them from sticking together.

+ The crispy sage can be prepared by putting on a flat plate lined with cling film, cook one minute interval until crispy.

甜點 | 4 人份

黑朱古力梳乎厘
Rich dark chocolate soufflé

材料

黑朱古力雪葩

可可碎	50 克
淡忌廉	55 毫升
全脂牛奶	185 毫升
砂糖	30 克
蜂蜜	4 克
奶粉	15 克

梳乎厘粉漿

牛奶	180 毫升
雲尼拿豆莢	半條
幼砂糖	30 克
粟粉	8 克
蛋黃	40 克約 2 個
75% 黑朱古力	150 克

打起蛋白

蛋白	120 克
幼砂糖	20 克

裝飾

糖霜	10 克

提示

+ 製作梳乎厘粉漿：下黑朱古力前，最好讓雲尼拿牛奶糊冷卻一會，否則可可脂會容易分離出來。

+ 加入打起蛋白前，一定要將梳乎厘粉漿攪至幼滑。

+ 加入打起蛋白時，不要過分攪拌混合物，否則梳乎厘在焗時不會脹起。

做法

黑朱古力雪葩

1. 可可碎放入已預熱至攝氏 180 度的焗爐內焗 3 分鐘。

2. 其他材料放入小鍋，邊攪拌邊煮至微滾，倒入可可碎，拌勻，浸泡一晚。

3. 將混合物過篩，轉至雪糕機打至幼滑。

梳乎厘粉漿

4. 刮出雲尼拿籽，拌入熱牛奶內，靜置 10 分鐘。

5. 糖、粟粉和蛋黃放進大碗內，拌勻，倒入雲尼拿牛奶拌勻，用中火煮至濃稠。

6. 趁熱下黑朱古力，拌勻。

7. 用保鮮紙蓋着放入雪櫃冷藏。

完成

8. 蛋白和糖一起打至企身，分三次加入梳乎厘粉漿內拌勻。

9. 細蛋糕模用牛油和麵粉抹勻，倒入梳乎厘粉漿，放入已注入熱水的焗盤內，用攝氏 180 度焗 20-25 分鐘。

10. 焗好後，灑上糖霜，立即伴與黑朱古力雪葩享用。

Rich dark chocolate soufflé

INGREDIENTS

COCOA NIBS SHERBET

- 50 g cacao nibs, roasted
- 55 ml whipping cream
- 185 ml full cream milk
- 30 g sugar
- 4 g honey
- 15 g milk powder

SOUFFLÉ BASE

- 180 ml milk
- 1/2 pc vanilla pod
- 30 g castor sugar
- 8 g corn starch
- 40 g (2 pcs) egg yolks
- 150 g 75% dark chocolate

MERINGUE

- 120 g egg white
- 20 g castor sugar

DUSTING

- 10 g icing sugar

COOKING TIPS

+ Let the pastry cream cool down a bit before putting the chocolate inside otherwise it will split so easily.

+ Beat the pastry cream until smooth before adding the meringue.

+ Don't over stir the soufflé mixture while combing the soufflé base and the meringue, otherwise the soufflé would be flat.

METHOD

CACAO NIBS SHERBET

1. Roasted the cocoa nibs in the pre-heated oven (180°C) for about 3 minutes.
2. In a sauce pan, combine everything and bring them to a simmer, stir constantly. And then add the roasted nibs to the milk mixture and let it infuse over night preferably.
3. Strain the mixture and pour the mixture into an ice cream machine and churn it until smooth.

TO MAKE THE SOUFFLÉ BASE

4. Scrape out the seeds from the vanilla pod and add to the hot milk, let it infuse for about 10 minutes.
5. Combine the sugar, corn starch and egg yolk in a bowl and mix well, pour the infused milk into the mixture. Stir to mix. Cook under medium heat until thick.
6. Add the chocolate in while the pastry is still hot, stir to mix.
7. Cool down the pastry covered with cling film in the fridge.

TO FINISH THE SOUFFLÉ

8. Whisk the egg whites with sugar until stiff, then fold it into the smooth pastry cream in 3 times.
9. Spoon the mixture into the buttered and floured ramekin dishes. Place them in a baking tray and pour in boiling water. Bake at the pre-heated oven at 180°C for about 20-25 minutes.
10. Sprinkle with icing sugar. Serve it right away with the cocoa nibs sherbet.

黑朱古力梳乎厘

189

雞湯 White chicken stock

約 3 公升份量 | Yields about 3 Liters

材料

雞殼 / 雞骨	2.5 公升
甘筍（去皮切塊）	250 克
白洋葱（切角）	200 克
西芹（去皮切段）	50 克
黑胡椒粒	2 克
新鮮 / 乾月桂葉	1 片
新鮮百里香	3-4 棵
凍水	份量要足夠蓋過材料

做法

準 備

1. 雞殼用凍水洗淨，放入湯煲內，注入凍水，水要蓋過雞殼兩吋。

2. 用中火煮至微滾，確保不要大滾，同時撇去油分和浮沫，下雜菜料和香料（月桂葉、百里香、黑胡椒粒）。

3. 上湯保持微滾再煮約 1 小時，並需要撇去油分和浮沫。

完 成

4. 煮上湯超過兩個小時後，關火，撇去湯面的油分，撈起材料，將湯用細網篩過濾。

5. 將整碗上湯用冰水浴冷卻，冷卻後放在雪櫃可保鮮數天，或分開幾份放在冰格內可儲存幾個月。

INGREDIENTS

- 2.5 kg chicken bones carcass
- 250 g carrots, peeled and cut into chunks
- 200 g white onion, cut into quarters
- 50 g celery, peeled and cut into strips
- 2 g whole black peppercorns
- 1 pc dried/fresh bay leaf
- 3-4 stems fresh thyme
- enough cold water to cover the ingredients

METHOD

TO START

1. Rinse the bones under cold water and place them into a suitable sized stock pot. Cover the bones with cold water by about 2 inches.

2. Turn the heat to medium and slowly bring the bones to a simmer, making sure it doesn't come a boil, keep skimming while it is simmering. Add the Mirepoix (mixed vegetables) and the aromatics (bay leaf, thyme and peppercorns).

3. Let the stock gently simmer for another hour or so, skimming the surface as needed.

TO FINISH

4. Once the stock has cooked for at least 2 hours, you can strain it. First, skim off as much fat as possible from the surface. Then gently remove the solids and discard. Finally strain the stock through a fine sieve.

5. Cool the stock over ice water. Once cool, it can be stored in the fridge for a few days or it can be portioned and frozen for a few months.

魚湯 Fish stock

約 3 公升份量 | Yields about 3 Liters

材料

雜魚（切件）	2 公斤
白酒	150 毫升
洋葱（切片）	400 克
韭葱（只取白色部份切件）	200 克
黑胡椒粒	2 克
新鮮百里香	3-4 棵
凍水	份量要足夠蓋過材料

INGREDIENTS

- 2 kg mixed fishes, cut into chunks
- 150 ml cooking white wine
- 400 g white onion, cut into slices
- 200 g leek white part only, cut into chunks
- 2 g whole black peppercorns
- 3-4 stems fresh thyme
- enough cold water to cover the ingredients

做法

準備

① 雜魚洗淨，去掉魚鰓和血，放在湯煲內，注入水和白酒。

② 用中火煮至微滾，確保不要大滾，同時要撇去油和泡沫，下雜菜料、黑胡椒粒和百里香。

完成

③ 上湯微滾約 30 分鐘，熄掉爐火，加蓋靜置約 15 分鐘，魚湯用密篩過濾。

④ 將整碗上湯用冰水浴冷卻，冷卻後放在雪櫃可保鮮數天，或分開幾份放在冰格內可儲存幾個月。

METHOD

TO START

① Rinse the fishes under cold water (remove the gills and blood) and place them into a suitable sized stock pot. Cover the fish with cold water and white wine.

② Turn the heat to medium and slowly bring it to a simmer, making sure it doesn't come a boil rapidly, keep skimming while it is simmering. Add the Mirepoix (mixed vegetables), the peppercorns and herbs.

TO FINISH

③ Allow the stock to simmer gently for about 30 minutes. Turn the heat off and cover the pot with a lid, let it steep for about 15 minutes. Carefully strain the stock through a fine sieve.

④ Cool the stock over ice water. Once cool, it can be stored in the fridge for a few days or it can be portioned and frozen for a few months.

Basic Recipe

蝦湯 Shrimp stock
約 3 公升份量 | Yields about 3 Liters

材料

連殼小蝦（切件）	2 公斤
橄欖油	50 毫升
白蘭地酒	80 毫升
橄欖油	20 毫升
甘筍（去皮切粒）	200 克
洋葱（切粒）	200 克
茴香或韭葱（取白色部份切粒）	200 克
番茄（切粒）	2 個
茄膏	30 克
黑胡椒粒	1/3 茶匙
新鮮羅勒	1-2 棵
凍水	份量要足夠蓋過材料

做法

準備

1. 蝦切成小件，瀝乾水分。
2. 用 50 毫升橄欖油拌勻蝦件，單層排在焗盤上，放入已預熱攝氏 200 度的焗爐焗約 30 分鐘，直至蝦乾身及金黃，期間要翻轉蝦件。
3. 在焗蝦的同時，湯煲下 20 毫升橄欖油，用中火煎雜菜料至金黃，下番茄煮至乾身，拌入茄膏。
4. 下香料和蝦入湯煲；將白蘭地酒注入曾焗蝦的盤內拌勻，將帶有蝦香的白蘭地酒倒入湯煲。
5. 下水蓋過所有材料，用中火煮至微滾，確保不要大滾，同時撇去油分和泡沫。

完成

6. 上湯煮約 30 分鐘，熄掉爐火，加蓋靜置約 15 分鐘，蝦湯用密篩過濾。
7. 將整碗上湯用冰水浴冷卻，冷卻後放在雪櫃可保鮮數天，或分開幾份放在冰格內可儲存幾個月。將整碗上湯用冰水浴冷卻，冷卻後放在雪櫃可保鮮數天，或分開幾份放在冰格內可儲存幾個月。

INGREDIENTS

- 2 kg small shrimps with shells, cut into pieces
- 50 ml olive oil, 80 ml brandy, 20 ml olive oil
- 200 g carrot, peeled and diced
- 200 g white onion, diced
- 200 g fennel or leek, white part diced
- 2 pcs tomatoes, diced
- 30 g tomato paste
- 1/3 tsp whole black peppercorns
- 1-2 stems fresh basil
- enough cold water to cover the ingredients

METHOD

TO START

1. Cut the shrimps into small pieces and strain them to remove excessive water.
2. Toss them with 50 ml of olive oil and arrange them in one layer on a baking tray. Roast them in the pre-heated oven at 200°C, turn them occasionally for even roasting, until dry and golden brown (about 30 minutes).
3. Meanwhile, sauté the mixed vegetables with 20 ml of olive oil in a large pot over medium heat until golden brown. Add the tomatoes and cook until dry, add the tomato paste and stir to mix.
4. Add the aromatics and the roasted shrimps to the pot, deglaze the roasting tray with brandy and add to the pot as well.
5. Add water to cover and turn the heat to medium and slowly bring it to a simmer, making sure it doesn't come a boil rapidly, keep skimming while it is simmering.

TO FINISH

6. Allow the stock to simmer gently for about 30 minutes. Turn the heat off and cover the pot with a lid, let it steep for about 15 minutes. Carefully strain the stock through a fine sieve.
7. Cool the stock over ice water. Once cool, it can be stored in the fridge for a few days or it can be portioned and frozen for a few months.

雜菜上湯 Vegetable stock

約 3 公升份量 | Yields about 3 Liters

材料

橄欖油	15 克
韭葱（切粒）	200 克
甘筍（去皮切粒）	200 克
洋葱（切粒）	400 克
茴香（切粒）	200 克
西芹（切粒）	200 克
蒜頭（切半）	1 個
百里香	5 克
意大利芫茜	5 棵
新鮮 / 乾月桂葉	3 片
水	份量要足夠蓋過材料

INGREDIENTS

- 15 g olive oil
- 200 g leek, diced
- 200 g carrot, peeled and diced
- 400 g white onion, diced
- 200 g fennel bulbs, diced
- 200 g celery stick, diced
- 1 pc whole garlic head, cut into halves
- 5 g thyme
- 5 sprigs parsley stem
- 3 pcs dried/fresh bay leaf
- enough cold water to cover the ingredients

做法

準備

1. 橄欖油倒入大鍋內，用中火加熱，下韭葱、洋葱、茴香、甘筍、西芹和蒜頭，轉至小火煮約 10 至 15 分鐘，途中不斷攪拌。

2. 下水和其餘材料煮滾，轉小火微滾 1 小時（不用加蓋）。

完成

3. 雜菜上湯冷卻至室溫，過篩，棄掉雜菜。

4. 放在雪櫃可保鮮數天，或分開幾份放在冰格內可儲存幾個月。

METHOD

TO START

1. Heat olive oil in a large saucepan over medium heat. Add leek, onion, fennel bulbs, carrot, celery and garlic; reduce heat to low, and cook for about 10 to 15 minutes, stirring frequently.

2. Add water and remaining ingredients, and bring to a boil. Reduce heat, and simmer, uncovered for about 1 hour.

TO FINISH

3. Cool it down to room temperature, strain the mixture through a sieve, and discard the vegetables.

4. It can be stored in the fridge for a few days or it can be portioned and frozen for a few months.

雲尼拿蛋奶醬
Crème Anglaise
約 300 毫升份量 | Yields about 300ML

材料

全脂牛奶	250 毫升
蛋黃	60 克
雲尼拿豆莢	1/2 條
幼砂糖	25 克

INGREDIENTS

- 250 ml full cream milk
- 60 g egg yolks
- 1/2 pc vanilla pod
- 25 g castor sugar

做法

準備

❶ 剝開雲呢拿豆莢，刮出雲尼拿籽。在小鍋內用中火加熱牛奶和雲尼拿籽，熄掉爐火靜置 4 至 5 分鐘。

❷ 在大碗內拌勻蛋黃和糖，將熱牛奶混合物逐少逐少倒入蛋漿內，否則蛋漿會煮熟。

❸ 將牛奶蛋漿拌勻後倒入小鍋內，用中火邊煮邊攪拌 4 至 5 分鐘，直至濃稠可在匙背留痕。

完成

❹ 用密篩將雲尼拿蛋奶醬過濾至碗內，冷卻後放在雪櫃可以儲存三天。

METHOD

TO START

❶ Split open the vanilla pod and scrape out the seeds. Heat the milk and vanilla seeds in a saucepan over a medium heat. Turn off the heat and let it infuse for about 4-5 minutes.

❷ Combine the egg yolks and sugar in a mixing bowl, pour the hot milk mixture into the egg yolk mix, a little at a time, so that the eggs do not start to cook.

❸ Return the mixture to the sauce pan over medium heat and stir continuously for 4-5 minutes, or until the mixture has thickened enough to coat the back of a spoon.

TO FINISH

❹ Strain the sauce through a fine sieve into a bowl, leave to cool and then refrigerate for up to 3 days.

番紅花水
Saffron water
約 100 毫升份量 | Yields about 100ML

材料

番紅花蕊 / 粉	1 克
水	100 毫升

- 1 g saffron strands/powder
- 100 ml water

做法

1. 水煮滾，加入番紅花，浸過夜。
2. 番紅花水用篩過濾至塑膠容器內，放入雪櫃可儲存兩星期。

METHOD

TO START

1. Boil the water and put the saffron strands/powder in, let it infuse overnight ideally.
2. Strain it into a plastic container and keep it in the fridge for up to 2 weeks.

牛油麵粉糊
Beurre Manie
約 100 克份量 | YIELDS ABOUT 100G

材料

軟無鹽牛油	50 克
麵粉（過篩）	50 克

INGREDIENTS

- 50 g unsalted butter, soften
- 50 g plain/cake flour, sieved

做法

1. 在大碗內拌勻牛油和麵粉。
2. 用保鮮紙將混合物捲成條狀，放在冰格可儲存兩個月。

METHOD

1. Combine them together in a mixing bowl, stir to mix.
2. Shape it into a sausage shape by using cling film, it can be kept in the freezer for up to 2 months.

Basic Recipe

195

焗脆羽衣甘藍 / 羅勒葉
Crispy baby kale/basil

10 片份量 | Yields about 10 pcs

材料

羽衣甘藍或羅勒葉（洗淨瀝乾） 10 塊
橄欖油 10 毫升
鹽、黑胡椒 適量

做法

❶ 羽衣甘藍或羅勒葉用橄欖油拌勻，碟上鋪上微波爐用保鮮紙，將葉倒轉放在碟上，下鹽和黑胡椒調味。

❷ 放入微波爐，用大火（1000W）煮 2 分鐘，將碟取出，並調整葉的形狀，再煮 2 分鐘。

❸ 如果不夠脆，可以放在焗盤上，放入已預熱攝氏 80 度的焗爐焗 30 分鐘，或直至爽脆。

❹ 放在密封容器內可以儲存 3 天，如果變軟可以再放入焗爐焗脆。

INGREDIENTS

- 10 pcs baby kale/basil leaves, cleaned and dried
- 10 ml cooking olive oil
- salt and pepper to judge

METHOD

❶ Rub the leaves with olive oil and arrange them upside down on a plate lined tightly with microwave-safe cling film, lightly season with salt and pepper.

❷ Cook them in the microwave oven at high powder (1000W) for about 2 minutes. Take the plate out and adjust the shape of the leaves and cook it again for another 2 minutes.

❸ If they are not crispy enough, they can be transferred onto a tray and put it into the pre-heated oven at around 80°C for about 30 minutes or until they are crispy.

❹ They can be kept in an air-tight container for up to 3 days. They can be crisped up again at the above temperature if they have gone soggy.

除了羅勒葉外，可用其他香草如法國芫茜 chervil、鼠尾草等。附圖是法國芫茜和羽衣甘藍。

Besides basil, you may also use other fresh herbs of your choice for this recipe, such as chervil or sage.
The image here shows chervil and baby kale.

基本食譜

儲存竅門
Storage Tips

香草
Herbs

羅勒・番茜（包括意大利及捲葉芫茜）

將羅勒如鮮花般紮成一束，修剪切口位，插入注了一吋水的玻璃瓶內，放在室溫就可以；切勿放入雪櫃，以免葉子變黑。這個方法可讓羅勒保鮮數天至一星期，羅勒並會長根，你可放在盆裏栽種。

此外剩下的羅勒也可以做青醬（pesto），舀進製冰粒格，可保鮮一個月。

細香葱・百里香・迷迭香・可食用鮮花

其他香草如細香葱、百里香、迷迭香等，可用微濕的廚房紙包裹，放入膠盒內，再放入雪櫃比較不凍的位置，如雪櫃門內。用廚房紙包裹香草時，勿緊緊紮實，要留些空位，以免濕氣令香草發霉。

此外，香草應臨採用時才清洗。

將可食用鮮花如三色紫羅蘭放入膠盒內，蓋上一張微濕的廚房紙，然後放入雪櫃，這可保鮮 3-4 天。

香草
Herbs

製作乾香草

也可製作乾香草。將已剁碎的羅勒或番茜、整棵的百里香或迷迭香放在碟上，置於涼爽、乾燥的位置數天，待香草乾透後放入密封瓶，然後放進雪櫃儲存。

棄去香草

新鮮的香草保鮮期不長，如葉子變黑或枯掉，莖有霉菌，都應該將香草棄掉。

上湯
Stocks

任何上湯都應放在冰格儲存，約可保鮮兩個月。本書內採用的雜菜上湯、雞湯、魚湯和蝦湯，可以製作大份量，然後注入製冰粒格內，可節省儲存空間；也可以將上湯抽真空、貼上標記，放在冰格可保鮮三個月。

蛋奶類製品
Egg and Dairy products

已開封的忌廉和牛奶一定要放在原盒內，並放入雪櫃儲存，宜在 4 天內使用，記着不要放入冰格，否則退冰時奶和水會分離，不能再用。剩下的蛋白如預計 3、4 天不會用，應放入冰格，可保鮮一個月；這些蛋白可用來做馬卡龍。至於蛋黃可放入雪櫃保鮮 1-2 天，但不要放入冰格。

糕餅粉糰或意大利麵糰
Pastry/pasta products

不論是甜或鹹的粉糰或千層酥皮，可將用剩的用保鮮紙包裹或抽真空，放入冰格可保存一個月。所以每次可做較大份量，用前一晚放在雪櫃普通格解凍即可。

Storage Tips

BASIL AND PARSLEY (ITALIAN AND CURLY)

A fresh bunch of basil can be treated like a bouquet of flowers: Just trim the ends, place in a glass with an inch or so of water, and place on the counter at room temperature. Don't keep it in the fridge, otherwise it will turn black easily. The basil will remain fresh for anywhere from a few days to a week. The basil would even grow roots after a few days in water, so you can plant it in flower pot. If you have any fresh basil left, you can also use them to make Pesto. Freeze it in an ice cube container for up to one month.

CHIVES, THYME, ROSEMARY AND FRESH EATABLE FLOWERS

Other herbs, like chives, thyme, and rosemary, require a slightly different approach. Wrap them loosely in damp kitchen paper and keep in a plastic container, and place them in the warmest part of the refrigerator; one of the compartments in the door works perfectly. Do not wrap the herbs tightly or the trapped moisture may cause them to mold prematurely; Do not rinse the herbs until just before using.

For the delicate eatable flowers like pansy which can be kept in a plastic container covered lightly with clean damp kitchen paper. In this way, it can be stored in a good condition in the fridge for up to 3-4 days.

HOW TO DRY FRESH HERBS

If you have more fresh herbs than you can use, dry them. Place the leaves on a plate (chopped if using basil or parsley; whole if using thyme or rosemary) and set aside in a cool, dry place for several days. Then store them in a resealable container in the refrigerator.

Tips

199

WHEN TO THROW FRESH HERBS AWAY

Fresh herbs are no longer fit to use and should be discarded when the leaves turn dark or brittle, or the stems begin to show traces of mold.

STOCKS

Any types of stock are freezer friendly, they can be kept in the freezer for up to 2 months in a container. The stocks are using in this book, like the vegetable stock, chicken stock, fish stock and shrimp stock can be prepared a big batch at one time and reduce it down to essence. After that, they can be frozen in ice cube container to save space. Ideally, they can be vacuumed, labeled and stacked up in the freezer for up to 3 months.

EGG AND DAIRY PRODUCTS

Opened cream and milk must be kept in the original box in the fridge no more than 4 days ideally, remember they can't be frozen otherwise they will separate or split and can't be used anymore once defrosted. If you have got some egg white left but won't use within 3-4 days, they can be kept in the freezer for up to one month; it is especially useful for making macaroons. On the other hand, egg yolk can only be kept in the fridge for only 1-2 days. Never freeze the egg yolk.

PASTRY/PASTA PRODUCTS

No matter sweet or savoury dough, if you have any left over, they can be nicely wrapped or vacuumed and kept in the freezer for up to one month. The same applies to the puff pastry, which takes so long to prepare so that you can make a bigger batch at one time. What you need to do is to defrost them in the fridge one night before.

儲
存
竅
門

烹飪詞彙
Glossary

B

Bake it blind（焗批皮）（動詞）

這是在沒有餡料的情況下，預先焗好批皮。做法是在批模鋪上批皮，再鋪上烘焙紙，加入焗爐用的石豆等重物，確保批皮在烘焙時不會變形。

如果需要批皮烘焙透，待批皮固定形狀後移去重物再焗，批皮才會變金黃。

Bisque（濃湯）（名詞）

通常指用焗龍蝦、蝦或雜菜做成的濃湯，味道濃郁，質感幼滑濃稠。

Blanch（白焓）（動詞）

將蔬菜或水果放在淡鹽水內滾指定時間，然後立即放在冰水或凍水喉下沖水冷卻，停止加熱煮熟。

C

Clarified butter（牛油清）（名詞）

牛油清由牛油提煉出來，煮溶牛油直至水、奶和脂肪完全分離，其脂肪便是牛油清。

牛油清的沸點是大約攝氏 252 度，比一般牛油的攝氏 160 至 190 度高，所以牛油清多用作高溫煎炸。

Curdling（凝固）（名詞）

這是一個做醬汁時不想見到的結果。用雞蛋做的醬汁如：雲尼拿蛋奶醬，因用過高溫度加熱而令雞黃煮熟，醬汁起粒狀。

凍食的醬汁如蛋黃醬，脂肪對蛋的比例太高也會發生凝固效果。

D

Deglaze（灒酒／上湯）（動詞）

將上湯或酒倒入曾焗或煎過肉類的鍋裏，溶解黏在鍋裏的啡色肉屑，用作煮醬汁。

F

Folding（捲勻）（名詞）

是指將兩種不同質感或密度的混合物輕力拌勻。混合物通常是已打發的，例如蛋白或淡忌廉。如果混合物是液體或是乾的，會用膠抹刀；如果是已打發的淡忌廉或蛋白，可以用鋼絲打蛋器。

G

Ganache（朱古力醬）（名詞）

是指用朱古力混合淡忌廉或牛油的醬，朱古力醬通常用作蛋糕、批或麵包的塗面、餡料或醬汁。

朱古力和忌廉的比例沒有一定規律，通常餡料的比例是 2 份朱古力對 1 份忌廉，而 1：1 的比例用於塗面。

Granita（冰沙）（名詞）

解釋看雪泥（sorbet）

Macerate（浸軟）（動詞）

利用酒類浸軟材料，增加材料的濕度和味道，通常會用作軟化乾果。

Meringue 馬令（名詞）

馬令是由蛋白和糖（幼砂糖或糖霜）打發而成，打發期間加入酸性物質如檸檬或塔塔粉。馬令可以製作蛋白脆餅、法式忌廉蛋白和火焰雪山。

馬令分成三種

+ 法式馬令：法式馬令能輕易在家中做到，只需要將幼秒糖和蛋白打至企身和有光澤，記着打蛋白時幼砂糖要逐少逐少加入。

+ 意大利式馬令：利用座枱攪拌機打起蛋白至濕性發泡，加入煲熱至攝氏 113-115 度的糖漿（糖＋水），攪勻便成。意大利式馬令比較穩定和適合直接進食。可趁熱加入魚膠凝固，用作不同批類的裝飾，如檸檬批。

+ 瑞士馬令：是煮熟或經過高溫消毒的馬令，拌勻蛋白和糖 （份量是 85 克蛋白加 100 克糖），坐入微滾熱水浴內，不斷攪拌直至溫度升至約攝氏 71 度，關火並不斷攪拌至濃稠企身並有光澤即成。這種馬令宜用來製作瑞士奶油忌廉。

Mirepoix（用來增香的植物）（名詞）

是指切碎的洋葱、甘筍和西芹的混合物，比例通常是 2 份洋葱、1 份甘筍、1 份西芹。可以是任何上湯、湯、燜菜和醬料的味道來源。

Pancetta（意大利鹽醃腩肉）（名詞）

跟煙肉差不多，兩者都是用生豬腩肉以鹽醃製而成，所以需要煮熟食。一般的煙肉會多一個製作程序，經過鹽醃後再以多種木材煙燻而成。

Parma ham/Prosciutto（巴馬火腿 /意大利火腿）（名詞）

兩者都是用豬腿肉或肩肉乾醃製（dry-cured）一段時間而成，以生吃為主。Prosciutto 在意大利文解作火腿，但巴馬火腿則是指在巴馬地區製成的火腿。

Poaching（浸煮）（名詞）

利用低溫的液體（約攝氏 70-80 度）煮熟材料，是比較健康的煮食方法，通常會將材料浸在牛奶、上湯或酒內。

Puree（蓉）（名詞）

通常是指蔬菜或水果蓉，將煮熟的蔬菜或水果用攪拌機打勻後過篩而成。

R

Reduce（減少）（動詞）

將湯、醬汁、酒或果汁持續煲滾，減少水分，從而增加材料的味道，煮時不需要加蓋，讓水蒸汽揮發。

Render（動詞）

加熱煮溶任何動物脂肪以作煮食用，例如豬、鴨脂肪；脂肪溶解後，要用密篩隔走任何雜質。

煮溶後的脂肪可以儲存在密封容器內 2 個月，放在冰格更可儲存 1 年。

To rest meat（靜置）（動詞）

肉類煮熟後移離焗爐或爐火，放在碟上，蓋上錫紙，靜置的時間取決於肉的大小，通常 5-15 分鐘已足夠。

靜置令肉汁不會流失，保持濕潤和肉味。

S

Sauté（煎）（動詞）

在鑊裏用少量的油或脂肪以高溫煎煮食材，食材通常會先切成小件或切片，從而縮短煮食時間。

當煎煮肉類和海鮮後，通常會加入上湯或酒溶解鑊內的肉屑，用作製作醬汁。

Searing（炙）（動詞）

以高溫快速將材料（通常是肉或魚類）表面炙成金黃，這是烹調肉類的重要技巧，能令肉類味道濃郁，同時也為外觀加分。

Sorbet/sherbet/granita 雪泥 / 雪芭 / 冰沙

+ Sorbet：是用水果、水、糖、酸性材料拌勻，再在雪糕機內攪拌而成的甜品，不含奶類。
+ Sherbet：它是雪泥的軟滑版。在雪泥的材料內加入淡忌廉、奶和蛋白，比普通的雪泥更加幼滑濃郁，但仍然比一般雪糕輕怡。
+ Granita：跟雪泥的材料一樣，但不需要用雪糕機，取而代之是將材料放在鍋裏拌勻，再放入冰格雪凍，期間取出刮幾次成薄片，它的口感比雪泥略為粗糙。

Simmer（煮）（動詞）

將材料放在冒出泡沫的微滾水內（約攝氏 90-95 度）煮熟。

Skim（撇去）（動詞）

是指在煲煮上湯或醬料等液體時，撇去湯面或醬面的脂肪及泡沫。

Steep（浸泡）（動詞）

將茶或香料等乾材料浸在液體內，茶或香料的味道會釋放於液體內。

Sweat（炒軟食材）（動詞）

用低溫油加鹽，慢慢將材料如洋葱的水分炒去，但要確保不要有丁點焦黃。

Terrine（凍批）（名詞）

是指一種法式醬糜，用切碎的材料和肉醬加以定型製成。

Bake it blind (verb)

To bake a tart or pie crust without the filing; it is accomplished by lining a parchment paper, then filled with baking beans or stones to ensure the crust retains its shape while baking.
If a fully baked crust is required, the weights needed to be removed before pre-baking is complete so as to achieve a browned crust.

Bisque (Noun)

It is a smooth, creamy, thick and strongly seasoned soup, and is commonly made from roasted lobster, shrimp or crayfish and vegetables.

Blanch (verb)

To cook vegetables or fruits in salted boiling water for a timed interval, and straight away plunge into iced water or place under cold running water to stop/shock the cooking process.

Clarified butter (noun)

It is purified butter or butter fat made from melting butter, where the milk solids and water are separated from the butter fat.
It has a higher smoke point (252℃) than regular butter (160-190℃), and is therefore preferred in some cooking applications like sautéing and pan frying.

Curdling (noun)

It is the undesirable result of overheating the egg base hot sauce like sauce crème anglaise (vanilla sauce), the egg yolk will be cooked and end up with grainy texture.

In cold sauces like mayonnaise, too large ratio of fat to egg may also cause curdling.

D

Deglaze (verb)

To remove and dissolve the browned food residue left on a pan to flavour sauces after a piece of meat is roasted or pan-fried. A liquid such as stock, a spirit or some wine is added to act as a solvent.

F

Folding (noun)

It is technique to combine two types mixture with different density. It is normally used for items where something has previously been whipped such as egg white or cream which has lower density.

Folding is usually done with a rubber spatula for liquid and dry ingredients; or with wire whisk for whipped cream and egg white.

G

Ganache (noun)

It is a mixture of chocolate of any kind, whipping cream and sometimes butter for making a glaze, sauce or filling of pastries.

The ratio of chocolate to cream is varied. Typically, 2 parts chocolate to 1 part cream are used for making the filling; while 1 to 1 is commonly used for making a glaze.

Granita (noun)

Refer to sorbet.

M

Macerate (verb)

To make something, especially dried fruits, soft by leaving them in its cooking liquor so as to make them juicer and tastier.

Meringue (noun)

It is a mixture of egg white and sugar (castor sugar or icing sugar) which have been whipped up with an addition of acidic ingredient such lemon or cream of tartar. It can be used for making desserts like Pavlova, floating island and baked Alaska.

THERE ARE 3 TYPES OF MERINGUE:

+ French Meringue: It is the easiest and simplest meringue to make by home cooks, it is made by beating normal white sugar (e.g. castor sugar) into egg whites until stiff and shiny. Make sure the sugar is added slowly while whipping.

+ Italian Meringue: It is made by pouring the boiling sugar syrup (castor sugar + water, bring to a boil until reached 113-115 °C) into the softly whipped egg whites while the free stand mixer is still running at its full speed. It is more stable and safer to use without cooking. Gelatin can be added while it is still hot to make it suitable for decoration on pies, like lemon pie, without deflation.

+ Swiss Meringue: It is a cooked or pasteurized meringue made by combining sugar and egg whites (e.g. 85 g egg whites + 100 g sugar) in a mixing bowl sitting over a pan of simmering water; keep stirring until it has reached to around 71°C , then take it off the heat and whisk it until thick and glossy. It can be used for making Swiss butter cream.

Mirepoix (noun)

It is a mixture of roughly or nicely cut onions, carrots and celery, the normal ratio is 2 parts of onions, 1 part carrots, and 1 part celery. It is the flavour base for making stocks, soups, stews and sauces.

P

Pancetta (noun)

It is similar to bacon, they are both typically made from pork belly and cured for a certain length of time. Both are also considered "raw" and need to be cooked before eating. But bacon takes things one step further, it is cold-smoked with a wide range of woods after it has been cured.

Parma ham/prosciutto (noun)

Both of them are ham made from pork leg or shoulder has been dry-cured for a curtain period of time, it is normally served uncooked. 'Prosciutto' means ham in Italian, but if a ham is called Parma ham, it must be produced in Parma region in Italy

Poaching (noun)

It is a moist-heat, low temperature (about 70-80°C) and healthy method of cooking technique that involves cooking by submerging food in a flavoured liquid like milk, stock or wine.

Puree (noun)

It is a cooked food, usually vegetables or fruits, that has been cooked, blended and sieved to the consistency of a soft creamy paste.

R

Reduce (verb)

To intensify the flavour of a liquid mixture such as a soup, sauce, wine or juice by simmering or boiling. It is done without a lid, enabling the vapor to escape from the mixture.

Render (verb)

To melt and clarify the hard animal fat like pork fat and duck fat for cooking purposes. In this method, the fat is slowly cooking until it melts and then strained of impurities form the cooking process.

The rendered fat can keep in an air-tight container without going rancid for up to 2 months and for a year if frozen.

To rest meat (verb)

To give the meat a rest, in other words, let the cooked meat sit out of the oven or off the stove before you cut it. Simply transfer the cooked meat to a clean plate and cover it with foil, let it sit for 5-15 minutes (depending on the size of the meat).

Resting meat can prevent too much meat juices from running out onto the cutting board or plate, leaving the meat dry, tough and with less flavour.

S

Sauté (verb)

To toss food with a small amount of oil or fat in a shallow pan over relatively high heat, the ingredients used for sautéing are usually cut into small pieces or thinly sliced to facilitate fast cooking.

When meat or seafood is sautéed in a pan, which is often deglazed by adding stock or wine to remove the residue to make a sauce.

Searing (verb)

To cook the surface of the food (usually meat or fish) at high temperature until a caramelized crust forms. It remains essential technique in cooking meat because it creates desirable flavours through Maillard reaction; and the appearance of the food is usually improved with a well-browned crust.

Sorbet/sherbet/granita

+ Sorbet: It is a mixture fruit, water, sugar and acid blended together and churned in an ice cream maker to set; no dairy is added to sorbet, which is how it is different from ice cream.
+ Sherbet: It is sorbet's creamier cousin. Dairy like cream, milk or egg white is added to a sorbet mixture, and the result is a frozen dessert that is richer and smoother than sorbet but still lighter than ice cream.
+ Granita: It starts with the same base as sorbet, but instead of churning it in an ice cream maker, the base is poured in a pan and placed in the freezer. The surface is scraped a few times as it freezes, creating icy flakes that are coarser in texture than sorbet.

Simmer (verb)

To cook foods in a liquid at or just below the boiling point (around 90-95 ℃), where the formation of bubbles has almost ceased.

Skim (verb)

To remove something solid such as fat and impurities from the surface of a liquid like stocks and sauces while boiling and reducing.

Steep (verb)

To allow dry ingredients like tea or spices to soak in a liquid until it takes on the flavour of the dry ingredients.

Sweat (verb)

To cook foods like onions with salt added over low heat to help draw moisture away, make sure that little or no browning takes place.

T

Terrine (noun)

It is referred to a cold French forcemeat loaf similar to pate, made from coarsely chopped ingredients.

Seasons' 四季嘗樂 Feasts

作者	Author
	Eric Poon

策劃編輯	Project Editor
	Catherine Tam

攝影	Photographer
	Image Union

美術設計	Designer
	Nora Chung

出版者 — Publisher

Forms Kitchen

香港鰂魚涌英皇道1065號
東達中心1305室

Room 1305, Eastern Centre, 1065 King's Road,
Quarry Bay, Hong Kong.

電話　Tel:　2564 7511
傳真　Fax:　2565 5539
電郵　Email: info@wanlibk.com
網址　Web Site: http://www.wanlibk.com
　　　　　　　 http://www.facebook.com/wanlibk

發行者 — Distributor

香港聯合書刊物流有限公司
香港新界大埔汀麗路36號
中華商務印刷大廈3字樓

SUP Publishing Logistics (HK) Ltd
3/F., C&C Building, 36 Ting Lai Road,
Tai Po, N.T., Hong Kong

電話　Tel:　2150 2100
傳真　Fax:　2407 3062
電郵　Email: info@suplogistics.com.hk

承印者 — Printer

Best Motion

出版日期 — Publishing Date

二零一八年一月第一次印刷
First print in January 2018

版權所有 · 不准翻印

All right reserved.
Copyright©2018 Wan Li Book Company Ltd
Published in Hong Kong by Forms Kitchen,
a division of Wan Li Book Company Limited.
ISBN 978-962-14-6347-0